Table of Contents

1 Chapter 1: Engine Basics

The heart of any interstellar vehicle is its engine, a complex system requiring diligent care and understanding. This chapter will equip you with the foundational knowledge necessary to maintain three vital components: the ignition system, the fuel injectors, and the combustion chamber. Mastering these systems is the first step towards becoming a proficient rocket repair technician.

Ignition systems are the spark plugs of interstellar travel, initiating the controlled combustion that propels a rocket. They aren't simple flint-and-steel setups, though. Modern interstellar engines utilize highly sophisticated methods, often employing high-energy lasers or precisely timed electrical discharges to ignite the fuel mixture. These systems must operate with absolute precision, delivering the initial burst of energy at the exact moment and with the correct intensity to ensure a smooth and efficient engine start. Understanding the intricacies of laser alignment, energy output calibration, and the diagnostic tools used to troubleshoot these systems is paramount for any aspiring rocket mechanic. Imagine a faulty laser failing to ignite the fuel; a delayed launch could mean a missed window for optimal trajectory, stranding a crew light-years from their destination.

Fuel injectors are responsible for delivering the precise mixture of propellants into the combustion chamber. This isn't as straightforward as filling a gas tank. Interstellar engines often use exotic fuels, sometimes in cryogenic states, requiring specialized injection systems capable of handling extreme temperatures and pressures. These injectors utilize complex arrays of pumps, valves, and sensors, working in perfect harmony to ensure the correct ratio of fuel and oxidizer is delivered at the optimal rate. Think of them as culinary masters, carefully measuring and combining ingredients for the perfect recipe - except in this case, the recipe is for controlled explosions that generate immense thrust. A clogged injector, a leaky valve, or a malfunctioning sensor can lead to uneven combustion, reduced engine efficiency, or worse, a catastrophic engine failure.

The combustion chamber is where the magic happens. Here, the carefully measured and injected fuel mixture is ignited, transforming chemical energy into the raw power needed for interstellar travel. The chamber itself is a marvel of engineering, designed to withstand the extreme temperatures and pressures generated during combustion. It's constructed using advanced materials capable of resisting melting and erosion, often incorporating intricate cooling systems to maintain structural integrity. Imagine the forces at play within this chamber: superheated gases expanding violently, pushing against the chamber walls with unimaginable force. Even microscopic flaws in the chamber lining can lead to cracks, leaks, or even complete structural failure, jeopardizing the entire mission.

Maintaining these three vital engine components requires a deep understanding of their interconnectedness. A malfunction in one system can cascade, affecting the others. For example, a faulty fuel injector can lead to incomplete combustion, causing pressure buildup in the chamber and potentially damaging the ignition system. Therefore, a holistic approach to engine maintenance is essential. Regular inspections, meticulous cleaning, precise calibrations, and the ability to diagnose and troubleshoot problems quickly are all crucial skills for a competent rocket repair technician. This isn't just about fixing broken parts; it's about ensuring the safe and efficient operation of the very heart of the spacecraft, the engine that carries humanity beyond the stars. A single overlooked detail, a minor miscalculation, could have devastating consequences in the unforgiving environment of space.

1.1 Ignition Systems

A rocket's ignition system is its heart, the spark that brings the behemoth to life. Understanding its intricacies is paramount for any aspiring rocket engineer. These systems aren't simple light switches; they are complex orchestrations of energy, precisely timed and controlled to unleash the colossal power necessary for interstellar travel. Think of them as the conductors of a powerful orchestra, bringing together different instruments - the fuel, the oxidizer, the combustion chamber - to create a symphony of controlled explosions.

At the core of most modern rocket ignition systems lies the igniter. This small but mighty device is responsible for initiating the combustion process. Different types of igniters exist, each with its own advantages and disadvantages. Pyrotechnic igniters, for example, use a small explosive charge to generate a burst of hot gas and flame. These are reliable and relatively simple, but once initiated, they cannot be stopped. Another common type is the electric igniter, which utilizes a high-voltage spark to ignite the propellant. These offer more control, allowing for precise timing and even the possibility of multiple restarts. Hypergolic ignition, a third option, employs propellants that ignite spontaneously upon contact, eliminating the need for a separate igniter altogether. However, these propellants are often toxic and require careful handling. Choosing the right igniter is a critical design decision, balancing reliability, control, and safety.

The igniter's role is to create a small, localized flame. This initial flame, however, is not sufficient to ignite the entire engine. It needs to be amplified, spread, and intensified. This is where the preburner comes into play. The preburner is a smaller combustion chamber, fed with a small amount of propellant. It takes the initial spark from the igniter and amplifies it into a powerful torch of hot gas. Think of it as a blacksmith's bellows, fanning a small ember into a roaring fire. The preburner's output is then channeled into the main combustion chamber, igniting the primary propellant mixture and initiating the main engine burn. The timing of these events is crucial. The preburner must be activated just before the main propellant valves open. Too early, and the flame might extinguish before the fuel arrives. Too late, and the unburnt propellant could accumulate, leading to a potentially dangerous overpressure situation. This precise orchestration requires sophisticated control systems. These systems monitor various parameters, such as pressure, temperature, and propellant flow, to ensure the ignition sequence proceeds flawlessly. They are the silent guardians of the engine, ensuring that each step happens at precisely the right moment.

Modern rocket ignition systems often incorporate redundancy. This means having backup igniters and preburners ready to take over if the primary system fails. Imagine a concert where a second conductor stands ready to step in should the first one fall ill. This redundancy is critical for mission success, ensuring that a single component failure doesn't jeopardize the entire launch. Redundancy also allows for multiple ignition attempts, increasing the chances of a successful engine start.

Beyond the igniter and preburner, other components play vital roles in the ignition process. Check valves, for instance, ensure that propellants flow in only one direction, preventing backflow that could lead to explosions. Filters remove impurities from the propellant, protecting the delicate components of the ignition system from damage. Pressure regulators maintain a consistent propellant flow, ensuring stable combustion. Each of these components, though seemingly small, is essential for the reliable operation of the entire ignition system.

Troubleshooting ignition problems requires a systematic approach. Start by checking the electrical connections to the igniter, ensuring they are secure and free of corrosion. Inspect the igniter itself for any signs of damage or wear. If the igniter is functioning correctly, move on to the preburner. Check the propellant lines for leaks or blockages. Verify that the preburner's valves are operating correctly. Finally, examine the control system, looking for any error codes or anomalies. A thorough understanding of the ignition system's components and their interactions is essential for effective troubleshooting.

The environment of space presents unique challenges for ignition systems. The extreme cold of space can affect the performance of some igniters and propellants. The vacuum can cause outgassing of materials, leading to contamination. Radiation can damage electronic components. To address these challenges, specialized designs and materials are used. For example, igniters may be heated before ignition to ensure proper function in the cold. Components are rigorously tested to ensure they can withstand the harsh conditions of space.

For aspiring rocket engineers, a deep understanding of ignition systems is essential. It's not just about knowing the names of the parts, but about comprehending the underlying principles that govern their operation. It's about understanding the delicate interplay of energy, pressure, and timing that transforms a dormant rocket into a roaring vehicle capable of reaching the stars. This knowledge is the key to designing, building, and maintaining the powerful engines that will propel humanity's future explorations of the cosmos. Mastering the art of ignition is about harnessing the very fire that drives our dreams of interstellar travel.

1.2 Fuel Injection

The intricate dance of propellants and controlled explosions within a rocket engine is orchestrated by the fuel injection system. Understanding this system is paramount for any aspiring rocket repair technician. Think of it as the heart of the engine, pumping lifeblood into the combustion chamber. It's a complex network of pipes, valves, pumps, and injectors, each playing a crucial role in delivering the precise mixture of fuel and oxidizer necessary for efficient combustion. A slight imbalance can lead to anything from reduced thrust to catastrophic engine failure. This chapter will equip you with the fundamental knowledge needed to diagnose, repair, and maintain these vital systems.

Fuel injection systems in interstellar rockets primarily utilize two main types of injectors: impinging and non-impinging. Impinging injectors, as the name suggests, involve streams of fuel and oxidizer colliding at high velocity. This collision creates a fine mist, promoting rapid and efficient mixing before entering the combustion chamber. These injectors are categorized further into like-on-like, where two streams of the same propellant impinge, and unlike-on-unlike, where fuel and oxidizer streams collide. Each type exhibits unique characteristics in terms of mixing efficiency and combustion stability, influencing the overall engine performance. Non-impinging injectors, on the other hand, introduce the propellants into the combustion chamber without direct collision. They rely on the turbulence and diffusion within the chamber to achieve mixing. While generally simpler in design, non-impinging injectors require careful consideration of chamber geometry and flow dynamics to ensure proper combustion.

The selection of injector type depends on various factors, including the specific propellants used, engine design, and desired performance characteristics. For example, cryogenic engines using liquid hydrogen and liquid oxygen often employ impinging injectors to ensure rapid vaporization and mixing of these extremely cold propellants. Engines utilizing hypergolic propellants, which ignite spontaneously upon contact, may utilize simpler non-impinging injectors. Understanding these nuances is crucial for diagnosing and troubleshooting fuel injection problems.

Moving beyond the injectors themselves, the pumps and valves that comprise the fuel delivery system are equally critical. Turbo pumps, often powered by a small pre-burner chamber, are responsible for delivering propellants at incredibly high pressures to the injectors. These pumps are marvels of engineering, operating at extreme speeds and temperatures. Their intricate design requires meticulous maintenance and careful inspection for wear and tear. Control valves regulate the flow of propellants, ensuring the correct mixture ratio and controlling the thrust output of the engine. These valves must operate with precision and reliability, withstanding the harsh conditions within the engine.

Troubleshooting a malfunctioning fuel injection system requires a systematic approach. Begin by carefully inspecting all visible components for any signs of damage, leaks, or blockages. Pay particular attention to the injector orifices, which can become clogged or eroded over time, affecting the spray pattern and mixing efficiency. Check for proper operation of the control valves, ensuring they are opening and closing fully and responding correctly to commands. Monitor the pressure readings throughout the system, looking for any deviations from normal operating parameters. Low pressure could indicate a pump malfunction, while high pressure might suggest a blockage downstream.

Remember, working with rocket propellants demands utmost caution. Many propellants are highly toxic, flammable, or corrosive. Always adhere to strict safety protocols, wearing appropriate protective gear and following established handling procedures. Proper ventilation is essential to prevent the buildup of hazardous vapors. Before working on any fuel injection system, ensure the system is completely inert and free of any residual propellants. Consult the engine's technical manuals for specific safety guidelines and decontamination procedures.

The next step in troubleshooting involves analyzing the engine's performance data. Telemetry from the engine control system can provide valuable insights into the health of the fuel injection system. Look for anomalies in chamber pressure, temperature, and thrust levels, which can point towards specific problems. For instance, fluctuating chamber pressure might indicate unstable combustion caused by improper propellant mixing. A drop in thrust could be a sign of reduced fuel flow due to a pump malfunction or a clogged injector.

Finally, specialized diagnostic tools can be employed to pinpoint the root cause of a problem. Ultrasonic flow meters can measure the flow rate of propellants through the pipes, revealing any restrictions or leaks. Endoscopes allow for visual inspection of internal components, such as injector orifices and valve internals, without disassembly. Pressure transducers provide precise pressure measurements at various points in the system, helping to identify the location of a blockage or leak. Mastering these diagnostic techniques is essential for becoming a proficient rocket repair technician. Remember, maintaining a rocket's fuel injection system is not merely a technical task; it's about ensuring the success of the mission and the safety of the crew. Your attention to detail and dedication to precision are paramount in this critical role.

1.3 Combustion Chamber

The heart of any rocket engine, the combustion chamber, is where the magic happens. Here, the carefully orchestrated mixture of fuel and oxidizer ignites, unleashing a controlled explosion that generates thrust. Understanding its intricacies is crucial for any aspiring rocket repair technician. The chamber's design isn't just about containing this explosive power; it's about directing and harnessing it efficiently to propel a spacecraft across the vast expanse of space.

Think of the combustion chamber as a highly specialized pressure cooker. It needs to withstand extreme temperatures and pressures while ensuring the complete combustion of the propellants. Incomplete combustion leads to wasted fuel and reduced thrust, a critical issue millions of miles from the nearest repair dock. Different engine designs utilize varying chamber shapes and injection systems, each optimized for specific propellants and performance goals. A cylindrical chamber, for example, offers a good balance between volume and surface area, while a bell-shaped nozzle efficiently accelerates the exhaust gases, maximizing thrust.

The injector, a critical component within the combustion chamber, introduces the propellants in a precise mixture and pattern. Like a finely tuned showerhead, it atomizes the fuel and oxidizer, creating a mist that maximizes surface area for efficient combustion. Injector design is a complex science, balancing factors like flow rate, droplet size, and distribution pattern to achieve optimal performance and prevent instabilities. Imagine trying to light a campfire with a single log versus a pile of kindling—the finer the fuel particles, the easier and more complete the burn.

The chamber walls are more than just a container; they are an integral part of the combustion process. They must withstand the intense heat generated by the burning propellants, often exceeding thousands of degrees. Ablative liners, made of materials that char and erode in a controlled manner, are commonly used. This sacrificial layer absorbs heat, protecting the underlying chamber structure and maintaining its integrity. Think of it like the heat shield on a spacecraft re-entering the atmosphere, sacrificing itself to protect the delicate instruments inside. Cooling channels within the chamber walls circulate coolant, typically the fuel itself, to further manage the extreme temperatures. This regenerative cooling system not only protects the chamber but also preheats the fuel, improving its combustion efficiency. It's a clever system of two birds with one stone, maximizing both performance and longevity. Imagine running cold water through pipes embedded in the walls of a furnace - the water keeps the furnace from melting, and the heat warms the water, ready for use.

Inspection and maintenance of the combustion chamber are paramount. Even minor cracks or erosion can compromise its structural integrity, leading to catastrophic failure. Regular checks for wear and tear, especially around the injector face and nozzle throat, are essential. Specialized tools, like borescopes and ultrasonic detectors, allow technicians to examine the chamber's interior without disassembly, identifying potential problems before they escalate. Think of it like a dentist using an x-ray to find cavities before they become painful.

Replacing a damaged combustion chamber is a major undertaking, requiring meticulous precision and specialized equipment. The process involves carefully disconnecting fuel lines, injectors, and the nozzle assembly. Once removed, the old chamber is replaced with a new one, ensuring a perfect seal and alignment. It's a delicate operation, akin to heart surgery for a rocket, requiring a steady hand and a deep understanding of the engine's intricate workings.

Beyond routine inspections and replacements, advancements in materials science are constantly pushing the boundaries of combustion chamber technology. New alloys and composites offer increased temperature resistance and strength, enabling higher chamber pressures and improved engine efficiency. Innovative cooling techniques, such as film cooling, where a thin layer of coolant is injected along the chamber walls, are also being developed. These advancements promise more powerful and efficient engines, paving the way for even more ambitious space exploration. It's a testament to human ingenuity, constantly striving to push the limits of what's possible.

Understanding the complexities of the combustion chamber isn't just about fixing broken engines; it's about grasping the fundamental principles that make space travel possible. It's about appreciating the delicate balance of power, precision, and engineering that propels humanity beyond the confines of our planet. For the young engineer, mastering the intricacies of this crucial component is a key step on the journey to becoming a true interstellar mechanic. The combustion chamber, after all, is the fiery heart that beats within every rocket, driving us towards the stars. It's a testament to human ingenuity and the relentless pursuit of exploration.

2 Chapter 2: Navigation

Navigating the vast expanse of space presents unique challenges. Unlike terrestrial travel with its readily available landmarks and established routes, interstellar voyages require a different approach. We must rely on the celestial bodies themselves, the very stars that litter the cosmic canvas, to guide us. This is where star charts become invaluable. These intricate maps of the cosmos detail the positions and relative brightnesses of stars, forming recognizable patterns called constellations. Think of them as cosmic signposts, each constellation a unique marker in the three-dimensional ocean of space. Learning to identify these celestial patterns and their relationship to each other is paramount for successful interstellar navigation. It allows you to pinpoint your current location, chart a course to your destination, and critically, ensures you can find your way back home.

Imagine holding a star chart, a complex web of lines connecting distant suns. These lines represent the angular distances between stars, creating a framework for understanding their spatial relationships. By measuring the angles between known stars and comparing them to your chart, you can triangulate your position within a given sector of space. This process, known as celestial navigation, is a fundamental skill for any aspiring interstellar engineer. It's a blend of art and science, requiring a keen eye, precise measurements, and a deep understanding of the celestial sphere. While sophisticated computer systems assist in modern spacecraft navigation, mastering the basics of star chart reading remains crucial for redundancy and in emergency situations.

Beyond simply identifying constellations, star charts also encode valuable information about individual stars. Details such as spectral class, luminosity, and distance are often included, providing context and further refining navigational precision. For instance, knowing the spectral class of a star helps determine its color and temperature, aiding in identification. Luminosity, a measure of a star's intrinsic brightness, can be used to calculate distances, supplementing the angular measurements obtained from the star chart. These data points contribute to a more comprehensive understanding of the surrounding space, allowing for more accurate and efficient course plotting.

Now, let's delve into a more advanced concept: warp drive theory. While still largely theoretical, the principles behind faster-than-light travel are fascinating and crucial for understanding the potential for long-distance interstellar journeys. The core idea behind warp drive is the manipulation of spacetime itself. Instead of propelling the spacecraft through space at superluminal speeds, which would violate the laws of physics as we currently understand them, the warp drive aims to warp the fabric of space around the spacecraft. Imagine stretching a sheet of fabric, then bringing two distant points closer together by folding the fabric. The spacecraft remains within its local spacetime bubble, while the space around it contracts in front and expands behind, effectively moving the spacecraft across vast distances without exceeding the speed of light within its own localized frame of reference. This warping of spacetime is theoretically achieved by creating a "warp bubble" around the spacecraft. This bubble is a region of distorted spacetime, generated by manipulating exotic matter with negative mass-energy density. This exotic matter, still hypothetical, would interact with spacetime in a way that allows for the creation of the warp bubble. The manipulation of this exotic matter requires immense energy densities, currently beyond our technological capabilities. However, the theoretical framework provides a tantalizing glimpse into the potential future of interstellar travel, where journeys to distant star systems could be accomplished in significantly less time than currently envisioned using conventional propulsion methods.

Current research focuses on the nature of this exotic matter and the energy requirements for generating and sustaining a stable warp bubble. Some theories propose using highly focused gravitational fields or other forms of energy manipulation to achieve the desired spacetime distortion. The challenges are substantial, encompassing everything from the fundamental physics of spacetime to the engineering of devices capable of generating and controlling such immense energies. While practical warp drive technology remains a distant dream, ongoing research and theoretical advancements continue to pave the way for its eventual realization. Understanding the basic principles of warp drive theory provides a crucial foundation for future engineers who may one day bring this science fiction concept into reality.

The pursuit of warp drive and the mastery of interstellar navigation are intricately linked. As we venture further into the cosmos, the ability to traverse vast distances efficiently becomes paramount. Star charts provide the essential framework for navigating within our local galactic neighborhood, while warp drive technology holds the promise of bridging the gaps between stars, opening up a universe of possibilities for exploration and discovery. The combined understanding of these two critical areas forms the core of interstellar navigation, enabling us to reach for the stars and unlock the secrets of the universe. Mastering these skills will not only propel our understanding of the cosmos but also shape the future of interstellar travel for generations to come.

The next stage in interstellar travel will necessitate not only proficiency in plotting courses through conventional means but also a fundamental comprehension of how warp fields interact with spacetime. This interplay is complex and requires a deep understanding of theoretical physics. As warp technology develops, navigating within a warped spacetime environment will become a new frontier in interstellar travel, demanding innovative approaches to charting courses and accounting for the distortions introduced by the warp bubble itself.

Consider the implications of a warp bubble encountering gravitational fields or other celestial phenomena. The interaction could potentially destabilize the bubble or alter its trajectory, requiring sophisticated navigational systems to compensate. Future engineers will need to develop new methods for mapping spacetime within and around the warp bubble, accounting for the dynamic changes induced by the warp field. This will likely involve advanced sensor technologies capable of detecting and interpreting the subtle shifts in spacetime caused by gravitational forces and other cosmic influences. Furthermore, new navigation algorithms will be required to process this information in real-time and adjust the warp field parameters accordingly, ensuring safe and accurate travel through warped spacetime.

Finally, the development of warp drive technology raises a host of new navigational challenges related to relativistic effects. As the spacecraft approaches faster-than-light speeds, time dilation becomes a significant factor. The time experienced within the warp bubble will differ from the time experienced outside, potentially leading to discrepancies in navigational calculations and communication delays. Future interstellar navigators will need to account for these relativistic effects to accurately plot courses and maintain communication with other spacecraft and ground stations. This will likely involve the development of relativistic navigation algorithms that incorporate time dilation into their calculations, ensuring precise positioning and synchronization across vast interstellar distances.

2.1 Star Charts

Navigating the vast expanse of space requires more than just a powerful engine; it demands a deep understanding of celestial mechanics and the ability to interpret the universe's roadmap: the star chart. These charts aren't simply picturesque representations of constellations; they are complex tools that encode positional information, stellar classifications, distances, and much more. Mastering their use is paramount for any aspiring interstellar engineer. Forget your GPS, out here, it's the ancient art of celestial navigation that will guide you.

Imagine holding a star chart. What you see is a two-dimensional representation of a three-dimensional reality. Think of it as flattening a globe to create a map. Distortions are inevitable, but with practice, you learn to compensate for them. Star charts use a coordinate system similar to latitude and longitude on Earth, but adapted for the celestial sphere. These coordinates, known as Right Ascension and Declination, pinpoint a star's location with remarkable precision. Right Ascension, analogous to longitude, is measured eastward along the celestial equator from a fixed point called the Vernal Equinox. Declination, similar to latitude, measures angular distance north or south of the celestial equator.

Different charts serve different purposes. Some focus on specific regions of the sky, providing detailed information on individual stars and deep-sky objects. Others offer a broader perspective, depicting the entire celestial sphere on a single sheet. Some are optimized for specific times of year, reflecting the Earth's changing position in its orbit. Choosing the right chart is crucial for successful navigation. Understanding the projection used in the chart is also key. Common projections include the cylindrical, conical, and azimuthal, each with its own strengths and weaknesses. Cylindrical projections preserve direction, making them useful for plotting courses, while conical projections maintain area accuracy, making them suitable for studying specific constellations. Azimuthal projections are excellent for mapping regions around a specific point, such as the celestial poles.

Beyond simply locating stars, charts provide a wealth of information encoded in symbols and abbreviations. A star's spectral type, indicating its temperature and luminosity, is often represented by a letter. Variable stars, whose brightness fluctuates over time, are marked with special symbols. Nebulae, star clusters, and galaxies are also depicted, adding to the richness of the chart. Learning this symbolic language unlocks a deeper understanding of the universe. Think of it like learning a new language - one whispered by the stars themselves. It's a language of light and distance, of celestial bodies moving through the vast expanse of space.

Using a star chart effectively requires more than just understanding its symbols. You must also account for your own position in space and the time of observation. The Earth's rotation and orbital motion constantly shift our perspective of the sky. To compensate for this, star charts often include tools like planispheres or computer software that adjust the display for your specific location and time. Imagine you're on a spaceship hurtling through the void. Your onboard computer projects a dynamic star chart, constantly updating to reflect your changing position. As you approach a destination, the chart zooms in, revealing finer details of the star system. You verify your position against known pulsars, using their rhythmic pulses like celestial lighthouses.

Star charts are dynamic tools, constantly being refined and updated. New discoveries, improved measurements, and advances in our understanding of the universe all contribute to their evolution. Keeping your charts current is essential for accurate navigation. Imagine discovering a new nebula, its glowing gases previously hidden from view. This new information would need to be incorporated into the star charts, enriching our understanding of the cosmos and potentially opening up new navigation routes. Think of the early explorers charting unknown territories on Earth. As they made new discoveries, their maps evolved, reflecting a more accurate representation of the world.

As an interstellar engineer, your ability to interpret and utilize star charts is paramount. They are the bridge between your ship and the vast cosmic ocean you traverse. They are the key to unlocking the secrets of the universe, one star at a time. So, study them diligently, practice your celestial navigation skills, and always remember to carry the latest editions. The universe awaits, and your journey begins with a single point on a star chart. You plot your course, engage the engines, and embark on a voyage to the stars, guided by the ancient wisdom encoded within the celestial maps. Out there, amidst the swirling nebulae and distant galaxies, your understanding of star charts will be your compass, your guide, and your lifeline. It's more than just navigation; it's an understanding of the very fabric of space and time.

2.2 Warp Drive Theory

The concept of warp drive, while firmly rooted in science fiction, has intrigued physicists and engineers for decades. It represents the tantalizing possibility of bypassing the universal speed limit—the speed of light—and achieving faster-than-light travel. This isn't merely about getting to distant stars quicker; it's about fundamentally altering our understanding of spacetime and the possibilities of exploration. While practical warp drive remains elusive, the theoretical framework provides valuable insights into the nature of space, time, and gravity.

One of the most prominent theoretical frameworks for warp drive comes from the Alcubierre metric, proposed by physicist Miguel Alcubierre in 1994. This metric describes a spacetime geometry where a spacecraft could effectively travel faster than light by warping the space around it. Imagine spacetime as a flexible fabric. The Alcubierre metric suggests creating a "warp bubble" around the spacecraft, compressing the space in front of it and expanding the space behind it. The spacecraft itself remains within a locally flat region of spacetime inside the bubble, unaffected by the extreme warping outside. It's like surfing a wave; the ship rides within the warp bubble, carried along by the distortion of spacetime.

A crucial aspect of the Alcubierre drive is the concept of exotic matter. This hypothetical substance possesses negative mass-energy density, a property unlike anything we observe in the everyday world. Exotic matter is theoretically necessary to create the negative pressure required to warp spacetime in the manner described by the Alcubierre metric. While negative mass-energy density has not been directly observed, some quantum phenomena suggest its theoretical possibility. Understanding the properties and potential sources of exotic matter is a major hurdle in the pursuit of warp drive technology. Without a concrete method for generating and manipulating this unusual substance, the Alcubierre drive remains firmly in the realm of theoretical physics.

Moving beyond the Alcubierre metric, other theoretical warp drive concepts explore alternative methods for manipulating spacetime. These include the Krasnikov tube, which proposes creating a permanent "tunnel" through spacetime, allowing for near-instantaneous travel between two points. Unlike the Alcubierre drive, the Krasnikov tube requires immense amounts of energy to create but doesn't necessitate exotic matter. Another concept, the Natario warp drive, seeks to reduce the energy requirements of warp travel by modifying the shape of the warp bubble. This approach aims to create a more efficient and potentially achievable warp drive configuration.

While the mathematical frameworks for warp drives exist, the engineering challenges are immense. One significant obstacle is the sheer energy required to create and maintain a warp bubble. Estimates suggest that the energy needed is equivalent to the total mass-energy of a planet or even a star. Harnessing and controlling this level of energy is beyond our current technological capabilities. Moreover, the potential stability of a warp bubble is a major concern. Some theoretical models suggest that warp bubbles could be inherently unstable, prone to collapse or other unforeseen consequences. Navigating and controlling a warp bubble, even if one could be created, presents significant challenges.

Despite these considerable challenges, the continued exploration of warp drive theory has value beyond the pursuit of faster-than-light travel. Studying these concepts pushes the boundaries of our understanding of general relativity, quantum mechanics, and the nature of spacetime itself. It encourages innovative thinking and prompts us to question the fundamental limits of physics. The quest for warp drive, even if ultimately unsuccessful, can lead to breakthroughs in other areas of science and technology, perhaps in ways we cannot yet imagine. The pursuit of seemingly impossible goals often yields unexpected and valuable discoveries along the way. The exploration of warp drive theory serves as a reminder that the seemingly impossible can sometimes inspire groundbreaking advancements in human knowledge.

3 Chapter 3: Hull Maintenance

The unforgiving vacuum of space presents a constant barrage of threats to a spacecraft's hull, from micrometeoroids to radiation. Maintaining the integrity of this outer shell is paramount, not just for aesthetics but for the very survival of the crew and mission. This chapter will equip you with the essential knowledge and techniques for handling a range of hull maintenance tasks.

First, let's discuss meteorite damage repair. Impacts, even small ones, can compromise the hull's layered structure, leading to potential depressurization or structural weakness. Assessment is the initial step. Identify the extent of the damage, classifying it as superficial, penetrating, or structural. Superficial damage, often just a cosmetic pitting, can be addressed with specialized sealant compounds, applied meticulously to prevent further erosion from the harsh space environment. Penetrating damage, which breaches multiple layers, requires patching. These patches, typically composed of a self-sealing polymer combined with a reinforcing mesh, are applied over the damaged area and bonded using a thermal adhesive activated by a focused heat applicator. For severe structural damage, involving extensive deformation or rupture, a more complex procedure involving replacement panels becomes necessary. These pre-fabricated sections are fitted and secured via specialized clamps and then welded into place using a low-power electron beam welder, minimizing heat stress on the surrounding structure.

Next, we move on to shield regeneration. Modern spacecraft employ energy shields to deflect the constant bombardment of micrometeoroids and radiation. These shields, generated by powerful electromagnetic fields, gradually deplete over time and require periodic regeneration. The process involves channeling energy from the ship's reactor through a series of conduits to specialized emitter arrays embedded within the hull. Monitoring the shield integrity is crucial, and specialized diagnostic tools can pinpoint areas of weakness or depletion. Regeneration typically involves a controlled energy surge directed to the affected areas, restoring the shield's protective capabilities. However, overcharging can lead to instability and even catastrophic failure, requiring careful calibration and monitoring.

Now, let's consider hull painting. While it may seem like a purely cosmetic concern, the spacecraft's outer coating plays a critical role in thermal regulation. Specialized paints with high reflectivity minimize heat absorption from stars, preventing overheating. Conversely, sections exposed to minimal stellar radiation are coated in darker, heat-absorbing paints to maintain optimal internal temperatures. Applying these coatings involves careful surface preparation, including cleaning and priming, followed by the application of multiple layers of specialized paint, each designed for specific wavelengths of radiation. This meticulous process ensures both thermal protection and long-lasting durability.

Pressure sealing is another critical aspect of hull maintenance. Maintaining a pressurized environment within the spacecraft is essential for crew survival. Seals, typically made from robust, flexible polymers, are strategically placed throughout the hull, particularly around windows, hatches, and docking ports. Regular inspection is vital for identifying signs of wear and tear, including cracks, abrasions, or chemical degradation. Damaged seals should be promptly replaced using specialized adhesive applicators. The area around the seal must be thoroughly cleaned and prepared to ensure a perfect bond, preventing leaks and maintaining the integrity of the pressurized environment.

Finally, we address window replacement. The spacecraft's windows, offering breathtaking views of the cosmos, are also vulnerable to damage. Specialized transparent aluminum alloys, offering incredible strength and transparency, are used for these windows. However, micrometeoroid impacts can still cause cracks or chips. Replacing a window involves carefully removing the damaged pane using specialized suction tools, avoiding any further stress on the surrounding hull structure. The new window is then fitted into the aperture and secured using a combination of high-strength adhesive and mechanical clamps. A final pressure test confirms the integrity of the seal.

Mastering these hull maintenance techniques is essential for any aspiring rocket repair engineer. The harsh environment of space demands meticulous attention to detail and a deep understanding of the principles of material science, thermal dynamics, and structural engineering. Remember, the integrity of the hull is the first line of defense against the perils of interstellar travel. A well-maintained hull ensures the safety of the crew, the success of the mission, and the continued exploration of the cosmos. This hands-on approach, coupled with the theoretical knowledge you've gained, will ensure you are well-equipped to handle the challenges of keeping your spacecraft safe and operational in the vastness of space. Your proficiency in these techniques will determine your effectiveness as a rocket repair engineer. This practical experience, combined with your theoretical understanding, will ensure you're ready to tackle any hull-related issue that arises during interstellar travel.

3.1 Meteorite Damage Repair

Space is a harsh mistress, and even the toughest spacecraft hulls occasionally succumb to the onslaught of micrometeoroids and orbital debris. These impacts, while often small, can compromise the structural integrity of your vessel and, in the worst-case scenarios, lead to catastrophic decompression. This chapter focuses on the practical steps required to repair this type of damage, ensuring your rocket remains spaceworthy. Assessment is the first crucial step. Carefully examine the impact site, noting the size, depth, and shape of the damage. A simple visual inspection often suffices for smaller craters, but larger impacts may require ultrasonic testing to determine the extent of internal damage. Document everything with detailed photographs and measurements for accurate patch fabrication and insurance purposes. Once the damage is assessed, you'll need to choose the appropriate repair material. For minor surface scratches and pitting, a specialized epoxy resin, designed for vacuum conditions and extreme temperature fluctuations, can be applied. This resin cures quickly, forming a strong, airtight seal. Deeper impacts require more robust solutions. Pre-fabricated patches, composed of a lightweight, high-strength alloy compatible with your hull's material, are generally preferred. These patches come in various sizes and shapes and can be easily trimmed to fit the damaged area precisely. In rare cases involving significant structural damage, you may encounter scenarios requiring the replacement of entire hull panels. This complex procedure involves cutting out the damaged section and welding in a new panel, ensuring a perfect seal to maintain pressure integrity.

Preparing the damaged area is just as important as the patch itself. Thoroughly clean the impact crater and surrounding area with a specialized solvent to remove any debris, oxides, or contaminants. This ensures optimal adhesion of the repair material. For deeper craters, use a grinding tool to create a smooth, beveled edge around the perimeter of the damage. This bevel provides a larger surface area for the patch to bond to, increasing the strength and durability of the repair. If you're dealing with a punctured hull, temporarily seal the breach from the interior with a specialized sealant tape before proceeding with the external repair. This will prevent further air loss and create a safer working environment. Remember, safety is paramount in space.

Applying the patch requires precision and care. For smaller repairs using epoxy resin, simply mix the two-part epoxy according to the manufacturer's instructions and apply it evenly to the cleaned impact crater. Ensure complete coverage and a smooth, feather-edged finish to minimize drag. Larger patches require a different approach. Apply a thin layer of structural adhesive to the back of the pre-fabricated patch and carefully align it over the damaged area. Secure the patch with clamps designed for use in zero-gravity environments, ensuring uniform pressure distribution across the patch surface. Allow the adhesive to cure completely according to the manufacturer's specifications. After the adhesive has cured, inspect the repair thoroughly. Check for any gaps or imperfections in the seal. Perform a pressure test on the repaired section to confirm its integrity. Apply a thin layer of thermal paint over the repaired area to restore the hull's thermal properties and protect the patch from further micrometeoroid impacts.

Beyond these standard repairs, consider preventative measures. Regularly inspect the hull for signs of damage, particularly after passing through debris fields or encountering micrometeoroid showers. Applying a thin layer of sacrificial coating to vulnerable areas can help mitigate the effects of minor impacts. This coating, designed to erode upon impact, absorbs the energy of the micrometeoroid, preventing it from reaching the underlying hull material. Finally, maintaining the integrity of your spacecraft's shielding system is crucial. A fully functional shield will deflect the majority of incoming debris, minimizing the risk of hull damage and reducing the frequency of repairs.

Advanced repair scenarios might involve specialized equipment, such as laser welders for larger breaches or robotic arms for accessing difficult-to-reach areas. For extremely large punctures or structural damage that compromises the integrity of the pressure vessel, you'll need to seek assistance from a specialized repair facility. These facilities possess the advanced equipment and expertise necessary to handle complex repairs, including the fabrication and installation of replacement hull sections. Never hesitate to call for help when a repair exceeds your capabilities or resources. The safety of your crew and the success of your mission depend on it. Remember, a well-maintained hull is your first line of defense against the unforgiving environment of space. By diligently following these repair procedures and practicing preventative maintenance, you can ensure the longevity and safety of your spacecraft on its interstellar journey.

3.2 Shield Regeneration

A spaceship's shield is its first line of defense against the relentless barrage of micrometeoroids, radiation, and other hazards of interstellar travel. Maintaining its integrity is not just a matter of convenience; it's a matter of survival. This chapter will equip you with the knowledge and practical skills to keep your ship's shields at peak performance, ensuring the safety of the crew and the integrity of the vessel. Shield regeneration isn't about simply flipping a switch; it's about understanding the complex interplay of energy flow, material science, and rapid diagnostics.

Shield systems are typically composed of layered energy fields, projected from generators strategically placed throughout the hull. These fields, while invisible to the naked eye, interact with incoming particles, dissipating their kinetic energy and preventing them from impacting the physical structure of the ship. Think of it like a highly sophisticated, invisible armor. However, repeated impacts, especially from larger particles, gradually deplete the shield's energy reserves, weakening its protective capabilities. This is where the principle of regeneration comes into play.

The core of any shield regeneration system is the energy conduit network. This intricate web of superconducting cables channels power from the ship's main reactor to the shield generators. Regular inspection of these conduits is crucial. Look for signs of wear and tear, such as micro-fractures, corrosion, or loose connections. These seemingly minor issues can significantly impede energy flow, slowing down the regeneration process and creating vulnerabilities in the shield. Use a specialized diagnostic tool, like a multi-spectrum energy flow analyzer, to pinpoint areas of reduced conductivity. This handheld device emits a range of energy frequencies and measures the response, allowing you to identify even the slightest impedance.

Once the energy conduits are confirmed to be functioning optimally, the next step is to assess the integrity of the shield emitters themselves. These specialized components are responsible for projecting the energy field that forms the shield. Over time, they can suffer from degradation due to the constant bombardment of high-energy particles. A telltale sign of emitter malfunction is an uneven shield strength, creating localized weak points. Use a shield integrity scanner to map the shield's energy density, identifying any areas where the field strength is below optimal levels.

Replacing a faulty emitter requires precision and care. First, de-energize the affected section of the shield. Then, using a magnetic grapple, carefully detach the faulty emitter from its housing. Ensure the replacement emitter is of the correct specifications and is compatible with the ship's systems. Align the new emitter with the housing and secure it using the magnetic grapple, ensuring a firm connection. Finally, re-energize the shield section and re-scan to confirm the restoration of uniform shield strength.

Beyond regular maintenance and component replacement, understanding the nuances of energy management can significantly enhance shield regeneration efficiency. During periods of intense bombardment, prioritizing power allocation to the shields can accelerate the regeneration process. Conversely, during calmer periods, diverting power to other ship systems can conserve energy while still maintaining adequate shield strength. Mastering this dynamic balance between protection and power conservation is essential for long-duration interstellar voyages.

Furthermore, the composition of the shield itself plays a critical role in its regenerative capabilities. Advanced shield systems utilize adaptive materials that can dynamically adjust their molecular structure in response to incoming threats. These materials, often composed of exotic alloys and energized nanoparticles, possess enhanced energy absorption and dissipation properties, contributing to faster regeneration rates. Keeping these materials in optimal condition requires periodic treatments with specialized solutions that replenish their energy-absorbing capacity and prevent structural degradation. Consult the ship's manufacturer specifications for the recommended treatment protocols and materials.

Finally, remember that shield regeneration is not an instantaneous process. Even with fully functional systems and optimal energy allocation, it takes time for the shield to replenish its energy reserves after sustaining damage. Therefore, anticipating potential threats and proactively strengthening the shield before entering hazardous regions is crucial. By understanding the principles of shield regeneration and applying these practical techniques, you can ensure the continued protection of your vessel and its crew, allowing for safe and successful interstellar travel. Don't underestimate the importance of regular checks and meticulous maintenance - they are the cornerstones of a robust defense against the unpredictable dangers of space. This proactive approach, combined with a deep understanding of the technology, is what separates a skilled engineer from a novice tinkerer.

3.3 Hull Painting

A spacecraft's hull isn't just a metal shell; it's the first line of defense against the unforgiving void of space. Micrometeoroids, radiation, and extreme temperature fluctuations constantly bombard its surface. Maintaining the hull's integrity is paramount, but so too is its finish. A proper paint job isn't merely cosmetic; it's a crucial element of thermal regulation and protection. This chapter will guide you through the intricacies of hull painting, ensuring your spacecraft remains both functional and aesthetically pleasing.

Before any paint touches the hull, meticulous preparation is key. Begin by thoroughly cleaning the surface. Specialized cleaning solutions are formulated to remove space dust, debris, and any residual radiation clinging to the hull. These solutions often utilize nano-scrubbers to dislodge microscopic particles embedded in the metallic lattice. After cleaning, a visual inspection is mandatory. Look for any micro-fractures or pitting. Even seemingly insignificant imperfections can compromise the paint's adhesion and lead to larger problems down the line. Address these issues with a specialized sealant before proceeding. This sealant, typically a polymer-based compound, flows into the imperfections, creating a smooth and uniform surface.

Once the sealant cures, the hull is ready for priming. Primer acts as an intermediary layer between the metal hull and the final paint coat. It enhances adhesion, providing a stable base for the paint to grip. The specific primer used will depend on the hull's composition. For example, titanium alloys require a different primer than aluminum composites. Apply the primer evenly, using long, sweeping strokes to avoid creating ridges or bubbles. Allow the primer to dry completely before proceeding to the next stage. This drying period is crucial for ensuring proper bonding and preventing premature paint failure.

Choosing the right paint is essential for long-term hull protection and performance. Spacecraft paint isn't your average household enamel. It's a highly specialized coating engineered to withstand extreme conditions. These paints often incorporate ceramic microspheres to reflect radiation and regulate temperature. The color itself plays a significant role in thermal management. Lighter colors reflect more solar radiation, keeping the spacecraft cooler, while darker colors absorb more radiation, potentially useful in colder regions of space. Consult the spacecraft's specifications to determine the optimal color and type of paint for your particular vessel.

Applying the paint requires precision and patience. Specialized spray applicators are used to ensure an even coat, minimizing the risk of streaks or drips. These applicators utilize electrostatic charges to attract the paint particles to the hull, resulting in a uniform and durable finish. Multiple thin coats are preferable to a single thick coat. Thin coats dry faster and more evenly, reducing the likelihood of imperfections. Allow each coat to dry completely before applying the next. This process can take several hours, depending on the paint and environmental conditions.

Post-painting inspection is as crucial as the preparation phase. Use a high-intensity light source to scrutinize the finished surface. Look for any imperfections, such as bubbles, streaks, or areas of uneven coverage. Address any issues promptly. Small imperfections can be corrected with spot touch-ups, while larger problems may necessitate repainting the affected area. Once the inspection is complete, apply a protective clear coat. This clear coat acts as a final barrier against micrometeoroids and radiation, further extending the life of the paint job.

Maintaining a spacecraft's hull is an ongoing process. Regular inspections and touch-ups are essential for preserving its integrity and functionality. Document every paint job meticulously, noting the type of paint, primer, and clear coat used. This documentation will prove invaluable for future maintenance and repairs. Remember, a well-maintained hull is not just about aesthetics; it's about ensuring the safety and longevity of your spacecraft. It's a testament to the meticulous care and attention to detail that define a true interstellar engineer. The hull, after all, is more than just a shell; it's the vessel's skin, its shield, and a symbol of its journey through the cosmos.

3.4 Pressure Sealing

A spaceship's hull is its first line of defense against the unforgiving vacuum of space. Micrometeoroids, radiation, and extreme temperature fluctuations constantly bombard the exterior, making a perfect seal absolutely essential for maintaining a safe and habitable environment inside. Pressure sealing isn't just about slapping on some sealant; it's a meticulous process requiring a deep understanding of the materials involved, the forces at play, and the specific design of the spacecraft. A single, microscopic breach can lead to a catastrophic loss of pressure, so precision and thoroughness are paramount.

The first step in any pressure sealing operation is identifying the leak. This often involves sophisticated diagnostic equipment, like ultrasonic detectors that can pinpoint even the tiniest escapes of pressurized air. Sometimes, a simple visual inspection, using a specialized dye penetrant that reveals cracks under UV light, will suffice. Once located, the nature of the damage dictates the repair strategy. Small punctures or cracks can often be addressed with specialized epoxy resins designed to adhere to the hull material and maintain flexibility under extreme temperature variations. These resins are often applied in layers, carefully curing each one before applying the next to ensure a robust and long-lasting seal. For larger breaches, a patch might be necessary.

Patching a hull breach in space presents unique challenges. Unlike a tire patch on Earth, a spaceship hull patch must withstand the vacuum of space on one side and the pressurized atmosphere on the other. This requires specialized patch materials, often composed of multiple layers, including a pressure-sensitive adhesive, a reinforcing mesh, and a sealant layer. Applying the patch correctly, ensuring a perfect seal around the edges, is crucial. This often involves a combination of specialized tools, like vacuum-rated clamps and heat applicators, to activate the adhesive and conform the patch to the hull's contours.

Beyond patching and sealing individual leaks, regular maintenance and preventative measures are essential for long-term hull integrity. This includes periodic inspections of all seals, particularly around windows, hatches, and docking ports. These areas are subjected to repeated stress and movement, making them particularly susceptible to leaks. Specialized sealant tapes and gaskets, designed to withstand these dynamic forces, are employed in these critical junctions. The tapes and gaskets must be inspected and replaced at regular intervals to maintain their effectiveness.

Understanding the different types of sealants and their appropriate applications is vital. Some sealants are designed for high-temperature environments, like those near the engine exhaust, while others are formulated for extreme cold. Using the wrong sealant can lead to premature failure and compromise the hull's integrity. The material of the hull itself also plays a significant role in sealant selection. Different hull materials have different coefficients of thermal expansion, meaning they expand and contract at different rates with temperature changes. The sealant must be compatible with these variations to prevent cracking and separation.

Beyond the immediate repair, proper documentation is essential. Every repair, no matter how small, must be meticulously recorded, detailing the location, type of damage, materials used, and the technician who performed the work. This creates a comprehensive maintenance history for the spacecraft, enabling engineers to track potential problem areas and predict future maintenance needs. This documentation is also crucial for safety inspections and certifications, ensuring the spacecraft remains fit for interstellar travel.

Furthermore, the environment in which the repair takes place adds another layer of complexity. Performing a pressure sealing operation in the vacuum of space requires specialized tools and procedures. Astronauts must wear pressurized suits, which restrict movement and dexterity. Tools must be tethered to prevent them from floating away. The lack of gravity also means that sealant and patch materials behave differently than they would on Earth, requiring specialized application techniques.

Proper surface preparation is key to a successful seal. Before applying any sealant or patch, the area must be thoroughly cleaned and any debris or damaged material removed. This often involves using specialized cleaning solutions and abrasive tools, taking care not to further damage the surrounding hull. For metallic hulls, a corrosion inhibitor might be applied to prevent further degradation. For composite hulls, a specialized primer may be needed to ensure proper adhesion of the sealant.

Finally, comprehensive training is indispensable for any aspiring rocket repair technician. This involves not just theoretical knowledge of materials and procedures, but also hands-on practice in simulated space environments. Working in zero gravity, with limited mobility and specialized tools, presents unique challenges that can only be mastered through experience. This training also emphasizes safety protocols, ensuring technicians can perform their duties effectively while minimizing the risks inherent in space travel. The future of interstellar travel rests on the meticulous work of skilled technicians who understand the vital importance of a perfectly sealed spacecraft.

3.5 Window Replacement

A spaceship's windows aren't just for sightseeing; they're crucial structural components that safeguard the crew from the vacuum of space and harmful radiation. They also provide natural light, a vital factor for crew morale and psychological well-being on long interstellar voyages. Understanding how to replace a damaged window is therefore a critical skill for any aspiring rocket repair technician. This process demands precision, careful execution, and adherence to strict safety protocols. Before starting any repair, ensure you are tethered securely to the ship's structure. The vacuum of space is unforgiving, and a single misstep could lead to disastrous consequences.

The first step in replacing a window is identifying the damage. Micrometeoroid impacts are common in space and can cause tiny cracks or chips. Larger impacts, however, can shatter the entire pane, requiring immediate replacement. Inspect the damaged area carefully using a specialized diagnostic tool, such as a sonic scanner, to assess the extent of the damage. This device emits high-frequency sound waves that can detect even the smallest fractures, providing a comprehensive picture of the window's integrity. Documenting the damage with detailed images is crucial for insurance claims and post-repair analysis. Remember to wear protective eyewear during the inspection process; microscopic glass shards can pose a significant hazard.

Once the damage is assessed, the next step is to depressurize the section of the ship containing the affected window. This is a critical safety precaution, as working on a pressurized window could lead to explosive decompression. Seal off the affected compartment using emergency bulkheads to isolate it from the rest of the ship. Once isolated, slowly vent the air from the compartment, monitoring the pressure gauge closely to ensure a controlled depressurization. A sudden drop in pressure can damage sensitive equipment and create hazardous conditions within the compartment.

With the compartment depressurized, you can now begin removing the damaged window. Specialized suction cups, designed for use in zero gravity, can be attached to the window's surface for controlled removal. These suction cups are powered by miniature vacuum pumps and provide a secure grip, even on large and heavy window panes. Slowly and carefully detach the window from its frame, taking care not to apply excessive force. Damaged windows can be brittle, and rough handling could cause further fracturing and the potential release of hazardous glass shards.

After the damaged window is removed, carefully clean the window frame to ensure a proper seal with the replacement. Use a specialized cleaning solution designed for space applications, ensuring it is compatible with the window frame material. Residual sealant from the previous window must be completely removed. Inspect the frame for any signs of damage, such as warping or corrosion, which could compromise the integrity of the new window seal. Even minor imperfections in the frame can lead to leaks, so meticulous attention to detail is essential.

Now, you can install the new window. The replacement window should be identical to the original, specifically designed for the spacecraft's make and model. These windows are typically composed of multiple layers of reinforced glass and transparent aluminum, offering superior strength and radiation protection. Carefully position the new window in the frame, ensuring proper alignment. A slight misalignment can create pressure points that weaken the seal. Once the window is in place, apply a specialized sealant around the perimeter of the frame. This sealant is designed to withstand the extreme temperature fluctuations and radiation exposure of space, ensuring a durable and airtight seal. After the sealant is applied, allow it to cure completely. The curing time will vary depending on the specific sealant used. Consult the manufacturer's instructions for precise curing times. Premature repressurization can compromise the seal, leading to leaks. Once the sealant has cured, gradually repressurize the compartment, monitoring the pressure gauge carefully. Any sudden pressure changes should be investigated immediately. After repressurization, conduct a thorough leak test using a sensitive pressure sensor. This test verifies the integrity of the new seal and ensures the compartment is ready for reintegration with the rest of the spacecraft. A successful window replacement is crucial for maintaining the safety and habitability of the spacecraft. Through careful preparation, meticulous execution, and adherence to safety protocols, you can ensure a successful repair and contribute to the success of any interstellar mission.

4 Chapter 4: Life Support

A spacecraft is more than just an engine and a hull; it's a self-contained ecosystem, a miniature world sustaining life amidst the void. Ensuring the well-being of the crew on long interstellar voyages requires meticulous attention to the life support systems. These interconnected systems work in harmony to provide breathable air, clean water, and a semblance of normal gravity, all essential for human survival in the hostile environment of space. Understanding their intricacies is crucial for any aspiring rocket repair technician.

Oxygen generation is the cornerstone of life support. While carrying compressed oxygen is feasible for short missions, long-duration interstellar travel necessitates a more sustainable approach. Electrolysis, a process that uses electricity to split water molecules into hydrogen and oxygen, is a common method. The generated oxygen is then circulated throughout the spacecraft's habitable compartments, maintaining a breathable atmosphere. Regular maintenance of the electrolysis system, including checking the electrodes and replacing filters, is crucial. Furthermore, monitoring oxygen levels and partial pressures is essential to ensure a safe and comfortable environment for the crew. Specialized sensors and alarms are installed throughout the spacecraft to detect leaks or other malfunctions promptly.

Water, equally vital for survival, is a precious resource in space. Unlike oxygen, which can be produced from readily available compounds, water must be conserved and recycled. Wastewater, including shower runoff, urine, and even moisture from exhaled breath, is collected and processed through a multi-stage purification system. This system typically involves filtration, distillation, and chemical treatment to remove impurities and contaminants. Regular maintenance of the water recycling system, including replacing filters and checking pumps and valves, is critical to prevent system failures and ensure a continuous supply of potable water. Redundancy in the system is also essential, with backup purification units ready to take over in case of primary system malfunction.

Beyond air and water, gravity plays a critical role in human health. Prolonged exposure to microgravity can lead to muscle atrophy, bone loss, and cardiovascular issues. To mitigate these effects, spacecraft designed for long-duration interstellar travel often incorporate artificial gravity systems. One common method involves rotating sections of the spacecraft, generating centrifugal force that mimics the effects of gravity. The speed of rotation and the radius of the rotating section determine the strength of the artificial gravity. Maintaining the rotational mechanisms, including bearings and motors, is crucial for the continued well-being of the crew. Regular inspections and lubrication are necessary to prevent mechanical failures and ensure smooth, consistent rotation.

Another critical aspect of gravity control is managing the Coriolis effect, a phenomenon caused by the rotation of the spacecraft. This effect can disorient crew members and affect the trajectory of moving objects within the rotating section. Understanding the Coriolis effect and its implications is essential for technicians working on the gravity control system. Specialized training and simulations are often employed to prepare technicians for working in a rotating environment.

Regular system checks and preventative maintenance are essential for the smooth functioning of the life support systems. Technicians need to be proficient in diagnosing and troubleshooting problems, from minor leaks to major system failures. They should be familiar with the schematics of the various systems and be able to identify and replace faulty components. Specialized diagnostic tools, including pressure sensors, flow meters, and gas analyzers, are used to monitor system performance and identify potential issues.

Furthermore, understanding the interplay between the different life support systems is crucial. A malfunction in one system can have cascading effects on others. For instance, a leak in the water recycling system could impact the oxygen generation system, leading to a decrease in breathable air. Therefore, technicians must be trained to assess the overall health of the life support ecosystem and identify potential points of failure. Regular simulations and emergency drills are essential to prepare the crew and technicians for any contingency. These exercises help to reinforce procedures and ensure a coordinated response in case of a real emergency. A well-trained and prepared crew is the best defense against the unpredictable challenges of interstellar travel.

4.1 Oxygen Systems

Oxygen generation is the cornerstone of any life support system, providing the breathable air necessary for survival in the vacuum of space. Understanding how these systems function, and more importantly, how to maintain them, is critical for any aspiring rocket repair technician. There are several approaches to oxygen generation, each with its own advantages and disadvantages. Electrolysis, for example, uses electricity to split water molecules into hydrogen and oxygen. This method is reliable and produces pure oxygen, but it requires a significant energy input. Another approach is chemical oxygen generation, using canisters that react to produce breathable air. These are easier to manage for short missions or as emergency backups, but the canisters require regular replacement and can pose a storage challenge on longer voyages. Solid oxide electrolysis cells (SOECs) offer a more efficient approach by using high temperatures to enhance the electrolysis process, reducing the energy requirement compared to traditional electrolysis. However, the high operating temperatures introduce material challenges and require specialized control systems.

Choosing the right system for a given spacecraft depends on mission duration, crew size, and available power. Once a system is chosen, regular maintenance is paramount. Electrolysis systems require periodic cleaning of the electrodes to maintain optimal performance. Membranes in the electrolysis cells can degrade over time and need replacement. For chemical oxygen generators, checking the expiration dates and pressure levels of the canisters is crucial. SOECs necessitate meticulous monitoring of the operating temperature and pressure, along with regular inspection of the seals and components for any signs of wear or damage. Beyond the generation process, oxygen delivery and regulation are equally important. Understanding the intricacies of the airflow within the spacecraft is essential.

Distribution systems typically consist of a network of ducts and vents designed to circulate the oxygen evenly throughout the living quarters and other areas of the ship. Maintaining proper airflow involves inspecting these ducts for leaks or blockages and ensuring the vents are functioning correctly. Oxygen sensors play a vital role in monitoring the oxygen concentration within the spacecraft, ensuring it remains within safe breathable limits. These sensors need regular calibration and replacement to ensure accurate readings. Furthermore, they are often connected to an automated control system that adjusts the oxygen flow based on the sensor data. Troubleshooting these control systems is a key skill for any repair technician, requiring a deep understanding of both the sensor technology and the control algorithms.

Another critical aspect of oxygen systems is the removal of carbon dioxide (CO2). As crew members exhale, CO2 builds up in the air, and if left unchecked, it can reach dangerous levels. CO2 scrubbers are essential components of the life support system, using chemical reactions to remove CO2 from the circulating air. These scrubbers require regular maintenance and replacement of the absorbent material. Monitoring the efficiency of the CO2 removal system is just as important as monitoring oxygen levels. Specialized sensors measure the CO2 concentration, and technicians must know how to interpret these readings and troubleshoot any issues.

Oxygen systems are also interconnected with other life support systems. For instance, water recycling systems often work in conjunction with electrolysis-based oxygen generation, providing the water needed for the process. Understanding these interdependencies is critical for comprehensive maintenance. A malfunction in the water recycling system can impact oxygen production, highlighting the importance of a holistic approach to life support system maintenance. Redundancy is a key design principle in spacecraft life support systems. Backup oxygen systems are essential in case of primary system failure. These backup systems often consist of emergency oxygen tanks or independent chemical oxygen generators. Regularly inspecting and testing these backup systems is vital to ensure they are ready to function in an emergency. This includes checking pressure levels, valve operation, and the integrity of the storage tanks.

Finally, understanding the unique challenges of oxygen systems in space is crucial. Microgravity can affect the distribution of oxygen within the spacecraft, leading to pockets of higher or lower concentrations. Radiation can also degrade materials used in oxygen systems, requiring specialized materials and shielding to mitigate these effects. The harsh vacuum of space presents another set of challenges, making leaks and pressure loss a constant concern. Thorough inspections and regular leak testing are essential for maintaining the integrity of the oxygen system and ensuring the safety of the crew. Recognizing these challenges requires specialized training and a deep understanding of the physics of space travel. This knowledge is vital for effective maintenance and troubleshooting, allowing technicians to keep the lifeblood of the spacecraft flowing smoothly.

4.2 Water Recycling

Water is the lifeblood of any long-duration space mission. Carrying enough water for a multi-year journey to another star system is simply not feasible given the mass constraints of interstellar travel. Therefore, efficient and reliable water recycling systems are absolutely critical for the survival of the crew. This chapter delves into the intricate processes involved in reclaiming and purifying every drop of water aboard your spacecraft, transforming wastewater into potable water suitable for drinking, hygiene, and hydroponic systems.

Understanding the various sources of wastewater is the first step in designing a robust recycling system. These sources can be broadly categorized into greywater, blackwater, and humidity condensate. Greywater originates from showers, handwashing, and laundry facilities. Blackwater, on the other hand, comes from toilets and contains a higher concentration of organic matter. Finally, humidity condensate, collected from the air conditioning system and other condensation points within the ship, represents a relatively clean source of recoverable water. Each of these sources requires a different approach to purification.

Greywater recycling typically involves a multi-stage filtration process. Initial filtration removes larger particles like hair and lint. This is followed by a finer filtration stage, often using ceramic or microfiltration membranes, to remove smaller suspended solids. Next, the water passes through an activated carbon filter to remove dissolved organic compounds and improve taste and odor. Finally, a disinfection stage, often utilizing UV light or ozone treatment, eliminates any remaining bacteria or viruses. This purified greywater can then be safely reused for non-potable purposes like toilet flushing and irrigation for onboard hydroponic gardens, significantly reducing the demand on potable water reserves.

Blackwater recycling presents a more complex challenge due to the higher levels of organic contaminants. The initial stages of blackwater processing often involve a membrane bioreactor (MBR). This combines biological treatment with membrane filtration. Microorganisms within the reactor break down the organic matter, while the membrane separates the treated water from the remaining sludge. The resulting effluent then undergoes further treatment similar to greywater recycling: filtration, activated carbon adsorption, and disinfection. While the idea of drinking recycled blackwater might seem unsettling, the rigorous purification process ensures the final product meets the highest standards of purity and is perfectly safe for consumption.

Humidity condensate, being the cleanest wastewater source, requires less intensive treatment. Often, simple filtration and disinfection are sufficient to render it potable. Collecting and utilizing this condensate not only contributes to the overall water supply but also helps control humidity levels within the spacecraft, preventing the growth of mold and mildew which can pose a health hazard to the crew and damage sensitive equipment. Monitoring the performance of your water recycling system is crucial. Regular checks of filter integrity, flow rates, and water quality indicators like pH, turbidity, and bacterial counts are essential. These checks allow for early detection of any issues, ensuring the continuous supply of clean water and preventing system failures.

In addition to these core recycling processes, advanced techniques like vapor compression distillation (VCD) are becoming increasingly important in long-duration space missions. VCD utilizes the heat generated from other spacecraft systems to evaporate water, leaving contaminants behind. The purified water vapor is then condensed, providing a highly efficient and reliable method for reclaiming water from various sources. Furthermore, research is ongoing into closed-loop hydroponic systems that integrate water recycling directly into the plant growth cycle, further minimizing water waste. These systems represent the future of sustainable life support in space.

Maintaining and troubleshooting your water recycling system requires a thorough understanding of its components and operation. Familiarize yourself with the various filters, pumps, membranes, and disinfection units. Learn how to replace filters, clean membranes, and diagnose common problems like reduced flow rates or unusual odors. Regular maintenance and preventative measures are essential to ensuring the long-term reliability of your life support system, safeguarding the crew's health and the success of the mission. Remember, in the vast expanse of space, every drop counts. Mastering the art of water recycling is not just a technical skill; it's a vital practice that ensures survival and sustains life far from Earth.

4.3 Gravity Control

Artificial gravity is not some futuristic fantasy; it's a critical system for any long-duration space voyage. Prolonged exposure to zero gravity leads to muscle atrophy, bone density loss, and cardiovascular issues. Maintaining a semblance of normal gravity helps mitigate these health risks and ensures the crew remains fit for duty. Currently, there are two primary methods for generating artificial gravity within a spacecraft: rotation and linear acceleration. Understanding the principles behind each is essential for any aspiring rocket repair technician.

Rotational gravity utilizes centrifugal force to mimic the effects of gravity. Imagine a large cylinder spinning in space. An astronaut standing inside, with their feet against the outer wall, would feel a force pushing them outwards. This outward force, generated by the rotation, acts as a substitute for gravity. The faster the cylinder spins, the greater the centrifugal force, and thus the stronger the simulated gravity. Repairing the mechanisms that drive and maintain this rotation is a crucial aspect of life support maintenance. These mechanisms include high-powered electric motors, bearings designed for continuous operation in a vacuum, and sophisticated control systems to regulate the rotation speed and ensure stability. Diagnosing faults within these systems requires a deep understanding of rotational dynamics and the ability to troubleshoot complex electromechanical systems.

Linear acceleration, the second method, involves constantly accelerating the spacecraft in a straight line. This continuous acceleration creates a force that pushes the occupants towards the back of the spacecraft, simulating gravity. While effective, this method requires a continuous expenditure of fuel, making it impractical for long journeys. However, understanding the principles of linear acceleration is vital as it plays a role in maneuvering and course correction. As a technician, you might encounter issues related to the inertial dampeners, crucial components that minimize the effects of sudden acceleration changes on the crew and equipment. These dampeners use complex magnetic fields to counteract the inertial forces, and their calibration and maintenance are critical for the safety and comfort of the crew.

Within rotational gravity systems, there are various designs, each with its own advantages and disadvantages. A large, ring-shaped structure spinning around a central hub offers a more stable and comfortable environment, minimizing the Coriolis effect, a phenomenon that can cause disorientation and nausea. However, such structures are complex to build and maintain. Smaller, modular designs using interconnected rotating pods can offer more flexibility and redundancy but may require more intricate control systems to manage their interconnected motion. Understanding these design variations is crucial for diagnosing and addressing specific mechanical failures. You might encounter issues with the rotational couplings between modules, imbalances in the rotating masses, or failures in the motors that drive the rotation. Each requires a different approach to repair, demanding a thorough understanding of the specific system at hand.

Beyond the core mechanisms of gravity generation, you'll also encounter supporting systems that contribute to the overall functionality and safety of the artificial gravity environment. These include sensors that constantly monitor the rotational speed and gravitational force, safety mechanisms that engage in case of malfunctions, and backup power systems to ensure continued operation in the event of a primary power failure. Understanding the interplay of these systems is critical for effectively troubleshooting and resolving issues. For example, a faulty sensor could trigger unnecessary safety shutdowns or lead to an imbalance in the rotation, causing discomfort and potential harm to the crew. Identifying the root cause of such problems requires a systematic approach and a comprehensive understanding of the entire gravity control system. Furthermore, maintaining these systems requires specialized tools and diagnostic equipment, including laser gyroscopes for precise measurement of rotation and high-precision force sensors to calibrate the artificial gravity levels.

Finally, understanding the human element of gravity control is essential. Regular checks of the crew's physical condition and monitoring their adaptation to the artificial gravity environment are crucial aspects of life support maintenance. This involves working closely with the medical team and understanding the physiological effects of prolonged space travel. By integrating technical expertise with a focus on crew wellbeing, you'll ensure the continued health and safety of those on board, contributing to the success of the mission. In the vast expanse of space, the smooth operation of the gravity control system is not just a technical matter; it's the foundation upon which the crew's health and mission success rely.

5 Chapter 5: Communication Systems

Deep space communication isn't as simple as picking up a phone. Consider the vast distances involved: signals can take hours, days, or even weeks to travel between starships and ground control. This time delay necessitates robust systems and a keen understanding of how to optimize them, especially when dealing with critical situations where a rapid response is essential. This chapter dives into the crucial aspects of interstellar communication, focusing on antenna alignment and signal boosting, two vital procedures for any aspiring rocket repair technician.

Antenna alignment is the bedrock of effective interstellar communication. These aren't the simple antennas found on terrestrial radios. We're talking about sophisticated, high-gain directional antennas designed to focus a signal towards a specific point in the vastness of space. Imagine shining a flashlight in a dark room. A focused beam illuminates a specific area, while scattered light barely reveals the surroundings. Similarly, a properly aligned antenna concentrates the transmitted signal, drastically increasing its effective range and minimizing power wastage. Misalignment, even by a fraction of a degree, can result in the signal missing its target entirely, rendering communication impossible. Therefore, understanding the principles of antenna alignment, and mastering the tools used to achieve it, is paramount for a rocket repair technician. This involves familiarizing yourself with the ship's navigation systems, using software to pinpoint the target's location, and employing fine-tuned mechanical adjustments to orient the antenna with absolute precision. Furthermore, factors like the ship's orientation and movement, as well as potential interference from celestial bodies, must be taken into account.

Maintaining precise alignment becomes even more critical when dealing with moving targets, like another spacecraft or a space station. Here, dynamic antenna tracking systems come into play. These systems automatically adjust the antenna's position in real time to maintain optimal signal strength, compensating for the relative motion between the communicating vessels. Understanding the intricacies of these tracking systems, their calibration, and potential points of failure, is essential for ensuring uninterrupted communication during critical maneuvers, such as docking or emergency rendezvous. Regular maintenance checks, including verifying the integrity of the tracking motors and sensors, are crucial to prevent communication blackouts when they are needed most.

Next, let's delve into signal boosting. Even with perfect antenna alignment, the sheer distance signals travel can weaken them significantly. This attenuation, coupled with interstellar noise and interference from cosmic radiation, can render a signal unintelligible. Signal boosting technologies address this issue by amplifying the transmitted signal and filtering out unwanted noise. One common method employs relay stations. Strategically positioned throughout space, these stations act as intermediaries, receiving weak signals, amplifying them, and retransmitting them towards their final destination. Think of them as stepping stones across a vast cosmic river, enabling signals to hop across the expanse of space. Maintaining these relay stations is an integral part of a technician's job, ensuring the smooth flow of communication across the interstellar network.

Another method of signal boosting involves advanced signal processing techniques. These techniques employ sophisticated algorithms to enhance the signal-to-noise ratio, effectively extracting the desired signal from the background noise. Imagine trying to hear a whisper in a crowded room. Signal processing is like having a device that filters out all the other voices, allowing you to clearly hear the whisper. These systems are often integrated directly into the communication hardware, requiring specialized knowledge to diagnose and repair. Familiarization with the intricacies of these algorithms and the hardware that implements them is essential for maintaining optimal communication performance. Technicians must be proficient in analyzing signal quality, identifying sources of interference, and adjusting the system parameters to maximize clarity and minimize data loss.

Finally, let's talk about the future of interstellar communication: quantum entanglement communication. While still largely theoretical, this technology promises instantaneous communication across vast distances by exploiting the peculiar interconnectedness of entangled particles. Imagine two coins flipped simultaneously, always landing on opposite sides regardless of the distance separating them. Quantum entanglement communication leverages a similar principle to transmit information instantaneously, bypassing the limitations imposed by the speed of light. Though still in its infancy, understanding the basic principles of quantum communication will become increasingly important for future rocket repair technicians, as this technology holds the key to revolutionizing interstellar communication and enabling real-time interaction across vast cosmic distances. Preparing for this future requires staying abreast of the latest developments in quantum physics and their applications in communication technology.

5.1 Antenna Alignment

Clear communication across the vast expanse of space relies heavily on the precise alignment of your spacecraft's antenna. Think of it like aiming a flashlight at a distant target - a slight misalignment can mean the difference between a bright, focused beam and a diffuse, ineffective glow. This chapter will equip you with the knowledge and practical steps to ensure your antenna is optimally positioned for crystal-clear interstellar communication. Understanding the principles behind antenna alignment, as well as the specific procedures for your spacecraft's model, is paramount for any aspiring rocket repair technician.

The first step is understanding the type of antenna your spacecraft utilizes. Parabolic dishes, helical antennas, and phased array antennas each have unique characteristics that influence their alignment procedures. Parabolic dishes, recognizable by their dish-like shape, concentrate signals towards a focal point, offering high gain and directionality. Helical antennas, resembling a coiled spring, are often used for circularly polarized signals, offering robustness against signal fading due to atmospheric effects. Phased array antennas, composed of multiple smaller antenna elements, allow for electronic beam steering, offering flexibility and the ability to track multiple targets. Identifying the type of antenna is the crucial first step towards proper alignment.

Once you've identified your antenna type, the next step is locating the antenna control panel. This panel will contain the necessary adjustments for azimuth and elevation. Azimuth refers to the horizontal angle of the antenna, while elevation refers to the vertical angle. These two angles, in conjunction, pinpoint the antenna's direction in three-dimensional space. Some spacecraft might incorporate more sophisticated systems, including automated alignment procedures or integrated software for precise targeting. Familiarizing yourself with the specific controls and indicators on your spacecraft's antenna control panel is essential.

Before beginning the alignment process, ensure you have the correct coordinates for your target. This information will typically be provided by mission control or stored within the spacecraft's navigation system. The coordinates will usually be expressed in terms of azimuth and elevation angles, providing the precise direction for optimal signal transmission and reception. Double-checking these coordinates is critical, as even a small error can lead to significant signal loss over interstellar distances.

With the target coordinates in hand, you can begin the alignment process. Start by adjusting the azimuth, slowly rotating the antenna horizontally until it's aligned with the target's azimuth coordinate. Then, adjust the elevation, carefully tilting the antenna vertically to match the target's elevation coordinate. These adjustments should be made smoothly and precisely, using the indicators on the control panel to confirm the antenna's position. Avoid jerky movements or forcing the antenna beyond its operational limits, as this could damage the delicate mechanisms.

Fine-tuning the alignment often involves utilizing a signal strength meter. This device measures the power of the received signal, allowing you to optimize the antenna's position for maximum signal strength. Slowly adjust the azimuth and elevation around the target coordinates, observing the signal strength meter for peak readings. The point of highest signal strength indicates the optimal alignment. This iterative process of adjustment and measurement ensures the most robust and reliable communication link.

Maintaining a stable connection often requires periodic adjustments. Factors like the spacecraft's movement, thermal expansion, and even gravitational forces can subtly alter the antenna's alignment over time. Regularly checking the signal strength and making minor adjustments as needed will ensure consistent communication quality. This proactive approach to maintenance prevents communication dropouts and ensures a stable connection throughout your interstellar journey.

Advanced alignment techniques may involve specialized equipment and software. For instance, some spacecraft employ beacon signals from known celestial objects for calibration. By aligning the antenna to these beacons, technicians can verify the accuracy of the onboard navigation system and fine-tune the antenna's positioning. These advanced methods provide a higher degree of precision, essential for critical communications and data transmission.

Troubleshooting antenna alignment issues requires a systematic approach. Begin by verifying the integrity of the antenna itself. Check for any physical damage, loose connections, or obstructions that could be interfering with signal transmission. Next, examine the cabling and connections between the antenna and the control panel, ensuring all connections are secure and free from corrosion. Finally, consult the spacecraft's technical manuals for specific troubleshooting procedures related to the antenna system. A methodical approach to problem-solving is essential for quickly identifying and resolving any alignment issues.

Beyond the technical aspects, understanding the broader context of antenna alignment is vital. Effective communication is the lifeline of any space mission, enabling vital data transmission, navigation updates, and emergency communication. A properly aligned antenna is the cornerstone of this communication network, ensuring the safety and success of the mission. By mastering the principles and practices of antenna alignment, you become a critical contributor to the exploration and understanding of the cosmos.

5.2 Signal Boosting

Deep space communication presents unique challenges. Distances are immense, signals weaken, and interference from celestial bodies can disrupt transmissions. Boosting signals is crucial for maintaining contact with mission control and other spacecraft. This chapter will equip you with the skills to enhance signal strength and ensure clear communication across the vastness of space. Understanding the principles behind signal amplification is the first step towards mastering this essential skill.

Amplification involves increasing the power of a signal. Imagine a ripple in a pond; a small pebble creates a small ripple, while a larger rock generates a bigger one. Similarly, a weak signal needs a "push" to travel further. This "push" comes from amplifiers, specialized devices designed to boost signal strength. Different types of amplifiers exist, each suited for specific applications. Solid-state amplifiers, for instance, are reliable and efficient, ideal for boosting routine communications. Traveling-wave tube amplifiers (TWTAs), on the other hand, offer higher power output, making them suitable for long-range transmissions and situations demanding exceptional signal strength. Knowing the strengths and weaknesses of each amplifier type allows you to choose the right tool for the job.

Beyond amplifiers, another crucial aspect of signal boosting involves focusing the signal. Think of a flashlight; a wide beam disperses the light, while a focused beam concentrates its power in a smaller area, increasing its intensity. Similarly, antennas play a crucial role in focusing and directing communication signals. Parabolic antennas, with their dish-like shape, are particularly effective at concentrating signals and increasing their effective radiated power (ERP). Precise antenna alignment is essential; even slight misalignments can significantly reduce signal strength and lead to communication failures. Regular maintenance and calibration of antenna systems are, therefore, crucial for reliable communication.

Optimizing signal transmission isn't just about boosting power; it's also about minimizing signal degradation. Space is filled with various forms of interference that can weaken or distort signals. Cosmic radiation, solar flares, and even the Earth's atmosphere can introduce noise and disrupt communication. Understanding these sources of interference helps you anticipate and mitigate their effects. Techniques like signal filtering and error correction coding play a vital role in cleaning up noisy signals and ensuring the integrity of transmitted data. Filters help remove unwanted frequencies, while error correction codes allow receivers to reconstruct the original message even if parts of the signal are corrupted.

Consider the specific communication needs of the mission. A short-range communication link between two nearby spacecraft might require only a modest signal boost, while communication with a distant probe exploring the outer reaches of the solar system demands far greater amplification and more sophisticated techniques. Factors such as distance, data rate, and the surrounding environment all play a role in determining the optimal signal boosting strategy. Adapting your approach based on these factors is critical for successful communication.

Troubleshooting communication issues requires a systematic approach. Start by checking the simplest things first, such as cable connections and power supplies. A loose cable or a faulty power source can easily disrupt communication. Next, verify antenna alignment and amplifier functionality. Specialized diagnostic tools can help pinpoint the source of the problem. Signal analyzers, for example, allow you to visualize the signal and identify any distortions or interference. Spectrum analyzers help you analyze the frequency spectrum and detect any unwanted signals. By systematically isolating and addressing each potential issue, you can effectively diagnose and resolve communication problems.

Redundancy is a key principle in spacecraft design, especially for critical systems like communication. Having backup systems ensures that communication can continue even if the primary system fails. Redundant amplifiers, antennas, and even entire communication channels provide a safety net in case of equipment malfunction or unexpected events. Regularly testing these backup systems is crucial to ensure they are ready to take over when needed.

Signal boosting in space requires more than just technical knowledge; it also demands careful planning and resource management. Power consumption is a major constraint in spacecraft operations, and amplifiers can be energy-intensive. Balancing the need for signal strength with the available power resources is a critical aspect of mission planning. Efficient power management strategies, such as using lower-power amplifiers for routine communications and reserving high-power amplifiers for critical transmissions, can help conserve energy and extend mission duration.

Maintaining clear communication is paramount for the success of any space mission. Signal boosting plays a critical role in overcoming the challenges of long-distance communication and ensuring that vital information flows smoothly between spacecraft and ground control. By understanding the principles of signal amplification, antenna technology, interference mitigation, and troubleshooting techniques, young engineers can become proficient in maintaining and enhancing communication links across the vast expanse of space. Mastering these skills is essential for enabling exploration, scientific discovery, and ultimately, the advancement of our understanding of the universe.

6 Chapter 6: Emergency Protocols

Space is unforgiving. When things go wrong, they often go wrong quickly and catastrophically. This chapter equips you with the knowledge to handle some of the most common, and most dangerous, emergencies you might encounter while working on a rocket. Understanding these procedures isn't just about fixing a problem; it's about survival. Escape pods are your last resort in a major emergency. Each pod is a self-contained lifeboat, equipped with its own life support, navigation, and communication systems. Locate the nearest pod access corridor. Inside, you'll find a control panel. Initialization involves a sequence specific to each pod model, detailed in the accompanying manual - familiarize yourself with it now, not when alarms are blaring. Once initiated, the pod will detach from the main vessel, using emergency thrusters to propel it to a safe distance. Note that escape pod trajectories are pre-programmed for the nearest habitable planet or space station; overriding these settings requires advanced navigational

knowledge and should only be attempted in dire circumstances, consulting the emergency navigation manual located within the pod itself.
Fire suppression systems are critical in the oxygen-rich environment of a spacecraft. Unlike fires on Earth, fires in space behave differently, burning faster and hotter. Identify the type of fire - electrical, chemical, or oxygen-fed - as this dictates the appropriate suppression method. Most modern rockets utilize inert gas flooding systems, which displace the oxygen and suffocate the flames. Locate the nearest manual release valve and activate it, directing the flow of suppressant towards the base of the fire. For smaller, localized fires, specialized fire blankets are available, designed to smother flames without depleting the ship's oxygen supply. Remember, your priority is containing the fire and preventing it from spreading to vital systems.
Radiation shielding is paramount during solar flares and when traversing high-radiation zones. Spacecraft hulls offer some protection, but additional measures are often necessary. Emergency radiation shelters are strategically located throughout the vessel, providing increased shielding. These shelters are equipped with monitoring systems to track radiation levels and environmental controls to maintain a stable atmosphere. Locate the nearest shelter and follow the entry procedures displayed on the access panel. Inside, you will find emergency supplies, including radiation sickness medication. Note: Remain within the shelter until the all-clear signal is given or radiation levels return to acceptable limits, constantly monitoring the internal readings.

First aid in space presents unique challenges. Microgravity affects blood flow and wound healing. Familiarize yourself with the ship's medical bay and its equipment. The medical kit contains specialized bandages, antiseptics, and medications designed for use in space. For severe injuries, a bio-regenerative unit can assist in stabilizing patients until they can reach a medical facility. Critically, understand the procedures for administering emergency injections and using the automated defibrillator. Every second counts in a medical emergency, and swift, informed action can be the difference between life and death.

Distress signals are a final plea for help when all other options have been exhausted. These signals transmit a coded message, including the ship's identification and location, to any nearby vessels or space stations. Activating the distress beacon is a two-step process, requiring both a physical key and a vocal confirmation code to prevent accidental triggering. The location of the key and the code are classified information, accessible only to authorized personnel. Once activated, the beacon transmits a continuous signal until deactivated or its power source is depleted. Understanding the activation procedure is critical, but remember, using the distress beacon should only be considered as an absolute last resort.

Remember, thorough preparation is your best defense against disaster. Familiarize yourself with these emergency protocols, practice them regularly, and ensure you know the location of all safety equipment. Your quick thinking and decisive actions could save not only your life, but the lives of everyone on board. The environment of space is harsh and unforgiving, demanding respect and meticulous preparation. By understanding these procedures, you are not just a repair technician, but a guardian of the ship and its crew.

6.1 Escape Pods

Escape pods represent the last line of defense in spacecraft emergencies. Understanding their operation is crucial for any young engineer. This isn't just about pushing a big red button; it involves a complex sequence of procedures designed to ensure the survival of the crew in the face of catastrophic failure. From initial detection of the emergency to the final touchdown on a safe planet or rendezvous with a rescue vessel, every step in the escape pod deployment process must be executed flawlessly. Think of it like a carefully choreographed dance between human action and automated systems, where precision and timing are everything.

The first critical element in any escape pod scenario is identifying the nature of the emergency. Is it a hull breach causing rapid decompression? A fire raging through critical systems? Or perhaps an imminent collision with an asteroid? The type of emergency dictates the urgency and specific procedures for escape pod deployment. For example, a slow leak in an oxygen tank requires a different response than a sudden explosion in the engine room. Identifying the problem quickly allows for the selection of the appropriate escape pod, as some may be better equipped for certain scenarios. Some pods might be optimized for atmospheric entry, while others prioritize long-duration survival in the void of space.

Once the emergency is assessed, the next step is initiating the escape pod sequence. This usually involves activating an emergency beacon and alerting the crew. Depending on the spacecraft design, activating the escape pod sequence might also automatically seal off sections of the ship, preventing the spread of fire or depressurization. Then comes the critical step of securing the escape pod, which involves closing hatches and ensuring all crew members designated for that particular pod are safely inside. In some spacecraft, individual escape pods are assigned to specific crew members based on their roles and responsibilities.

Inside the escape pod, another layer of complexity awaits. Life support systems must be activated, including oxygen generation, temperature regulation, and waste management. Navigation systems play a vital role, charting a course to the nearest safe haven or sending out distress signals to pinpoint the pod's location for rescuers. Depending on the sophistication of the spacecraft and the nature of the emergency, the escape pod might be equipped with basic supplies like food, water, and medical kits, extending the survival time of the occupants until rescue arrives. These supplies are carefully selected and stored to maximize their shelf life and minimize the space they occupy within the limited confines of the escape pod.

Furthermore, the escape pod's trajectory and landing sequence are meticulously planned. For atmospheric entry, heat shields protect the occupants from the intense friction generated during descent. Retro rockets fire at precise moments to slow the pod's velocity and ensure a safe landing. Impact cushioning systems within the pod further mitigate the force of landing, protecting the crew from injury. In scenarios involving a water landing, the escape pod might be equipped with flotation devices and emergency beacons to aid in recovery.

Finally, post-landing procedures are just as critical as the escape itself. Crew members must know how to communicate their location, access survival supplies, and administer basic first aid if necessary. Training for these procedures is essential, as clear thinking and quick action are paramount in survival situations. Regular drills and simulations ensure that crew members are familiar with every step of the escape process, from initial alarm to eventual recovery. This training can mean the difference between life and death in a real emergency. Remember, in the vast expanse of space, the escape pod is more than just a machine; it's a lifeline.

6.2 Fire Suppression

Fire suppression in the unforgiving vacuum of space requires a vastly different approach than on Earth. Forget water hoses and fire extinguishers; those are practically useless in a pressurized spacecraft. Instead, we rely on specialized systems designed to combat flames swiftly and with minimal impact on the ship's atmosphere. Understanding these systems is crucial for any aspiring rocket repair engineer. This knowledge could quite literally save lives.

First, let's delve into the detection aspect. Smoke detectors, while still employed, are less effective in zero-gravity where smoke doesn't rise in a predictable plume. Instead, ionization detectors, which sense the charged particles produced by fire, are more prevalent. These detectors are strategically placed throughout the spacecraft, providing comprehensive coverage and quicker response times. Furthermore, sophisticated monitoring systems constantly analyze the ship's atmospheric composition, looking for subtle shifts in gas ratios that might indicate a nascent fire, even before visible flames or smoke appear.

Next, consider the suppression agents themselves. Water, as mentioned earlier, is largely impractical. Deployed in zero-gravity, it would simply disperse into floating globules, potentially short-circuiting electrical systems and doing little to quench the flames. Instead, we primarily use halon replacements like HFC-227ea and inert gases such as nitrogen, argon, and carbon dioxide. These agents work by either disrupting the chemical chain reaction of the fire or by displacing the oxygen necessary for combustion. They're stored in pressurized tanks and released through strategically placed nozzles. Crucially, these agents are designed to be relatively harmless to humans in the concentrations used for fire suppression, a critical consideration in a sealed environment.

Delivery systems vary depending on the specific area of the spacecraft. In crew compartments, localized suppression systems using nozzles and carefully directed gas flows are favored. These systems aim to contain and extinguish the fire rapidly while minimizing the spread of the suppression agent throughout the entire ship. In less populated areas, such as cargo bays or engine rooms, total flooding systems might be employed. These systems rapidly fill the entire compartment with suppressant gas, ensuring complete extinguishment, even in hard-to-reach areas. Understanding the nuances of these different systems is paramount.

Imagine a scenario where a fire erupts in the engine room. The intense heat could potentially compromise the structural integrity of the spacecraft. In this case, a total flooding system, triggered automatically by the detection system, would inject inert gas into the engine room, quickly suppressing the fire and preventing further damage. Conversely, if a small fire were detected in a crew cabin, a localized system would target the specific area, minimizing disruption to the crew and preserving precious oxygen supplies.

Maintenance of these fire suppression systems is just as vital as understanding their operation. Regular inspections are crucial. Pressure checks ensure the suppressant tanks are adequately charged. Nozzle inspections verify they're clear of obstructions. Control panel tests confirm the system's responsiveness. Furthermore, the entire system undergoes periodic simulations, allowing engineers to evaluate its effectiveness and identify any potential weaknesses. These rigorous maintenance procedures are essential for guaranteeing the system's reliability in the event of an actual emergency.

Beyond the automated systems, manual intervention is sometimes necessary. Crew members are trained in fire suppression protocols, including the use of portable fire suppression units containing specialized agents. These units are designed for tackling small, localized fires that might not trigger the automated system, providing an additional layer of safety. Understanding how these portable units function and their limitations is essential for every crew member.

Understanding the composition of the ship's atmosphere is also critical for effective fire suppression. Different atmospheres require different suppressant agents and concentrations. For example, a spacecraft with a higher oxygen concentration might require a more potent suppression agent or a larger quantity of inert gas. Knowing the specific requirements for the spacecraft you're working on is essential.

Finally, consider the post-fire procedures. Once a fire is extinguished, the affected area must be thoroughly inspected for damage. Any compromised equipment must be repaired or replaced. The suppressant gas must be vented, and the atmosphere replenished. Understanding these post-fire protocols is crucial for restoring the spacecraft to a safe and operational state.

Mastering the intricacies of fire suppression in space is not just about understanding the technical details of the systems; it's about developing a comprehensive approach to fire safety. It's about combining technical knowledge with practical skills and a deep understanding of the unique challenges posed by the space environment. This holistic approach is what distinguishes a competent rocket repair engineer from a truly exceptional one. Remember, in the vast emptiness of space, your ability to handle emergencies effectively can be the difference between mission success and catastrophic failure.

6.3 Radiation Shielding

Space radiation presents a significant hazard to both crew and equipment during interstellar travel. Understanding its various forms and implementing appropriate shielding strategies is crucial for mission success. The three primary types of radiation encountered in space are Galactic Cosmic Rays (GCRs), Solar Energetic Particles (SEPs), and trapped radiation belts surrounding planets. GCRs originate from outside our solar system and consist of high-energy atomic nuclei. SEPs, on the other hand, are ejected from the Sun during solar flares and coronal mass ejections. Trapped radiation belts, like the Van Allen belts around Earth, consist of charged particles trapped by a planet's magnetic field. Each type poses unique challenges and necessitates different shielding approaches.

Shielding spacecraft from radiation involves a multi-layered approach, combining passive and active measures. Passive shielding relies on the inherent ability of materials to absorb radiation. The effectiveness of a material depends on factors like its atomic number, density, and thickness. High-density materials like lead, tungsten, and even water offer good protection against GCRs and SEPs by slowing down and absorbing the energetic particles. Hydrogen-rich materials, such as polyethylene, are particularly effective against high-energy protons found in SEPs and trapped radiation belts. The thickness of the shielding material determines how much radiation is absorbed; thicker shields provide better protection but also add to the spacecraft's mass, which impacts fuel efficiency and maneuverability. Therefore, careful optimization is necessary to balance protection with weight considerations.

Active shielding, a more advanced technique, uses magnetic or electric fields to deflect charged particles away from the spacecraft. While still under development, these systems offer the potential for lighter and more efficient shielding compared to passive methods. Electrostatic shielding creates an electric field around the spacecraft, repelling incoming charged particles. Magnetic shielding utilizes a strong magnetic field, similar to Earth's magnetosphere, to deflect charged particles away from the crew and sensitive equipment. These systems require significant power and present complex engineering challenges, but they hold promise for future long-duration missions.

For crewed missions, spacecraft design incorporates radiation-hardened components and specific shielding zones within the vessel. Habitation modules often include "storm shelters" with increased shielding to protect the crew during intense solar flares. These shelters typically utilize a combination of passive shielding materials and strategically placed water tanks, which offer excellent radiation protection. Radiation monitoring systems throughout the spacecraft provide real-time data on radiation levels, enabling crew members to take appropriate precautions when necessary. Mission planners also consider flight trajectories to minimize exposure to known radiation belts and utilize predictive models to forecast solar activity and adjust mission timelines accordingly.

Beyond the structural elements of the spacecraft, personal protective equipment (PPE) plays a crucial role in mitigating radiation exposure for astronauts during extravehicular activities (EVAs). Specialized spacesuits incorporate layers of radiation-resistant materials to provide added protection against GCRs and SEPs. Dosimeters worn by astronauts track their cumulative radiation exposure throughout the mission, allowing for personalized health monitoring and risk assessment. Medical countermeasures, including pharmaceutical interventions, are also being researched to mitigate the long-term health effects of space radiation.

Understanding the specific radiation environment of the mission is crucial for selecting the appropriate shielding strategy. Missions within Earth's magnetosphere primarily contend with trapped radiation belts, while interplanetary voyages face the constant bombardment of GCRs and the intermittent threat of SEPs. For journeys beyond the solar system, the radiation environment becomes even more complex, requiring advanced shielding techniques and potentially incorporating novel materials. Research into advanced materials, including composites and

nano-structured materials, continues to explore lighter and more efficient shielding solutions.
The principles of radiation protection extend to the maintenance and repair of spacecraft. Repairing damaged shielding components requires specialized procedures to ensure the integrity of the radiation protection system. Technicians working on the exterior of the spacecraft must follow strict safety protocols, including wearing appropriate PPE and minimizing exposure time in high-radiation areas. Repair kits may contain specialized patching materials and sealants designed for use in radiation environments. Regular inspections and maintenance of shielding systems are essential to detect and address any potential vulnerabilities. Understanding the types of radiation, the shielding materials used, and the repair procedures specific to the spacecraft are essential for maintaining a safe and effective radiation protection system throughout the mission.

6.4 First Aid

Space is a hostile environment. Even inside the carefully controlled environment of your spacecraft, accidents can and do happen. This chapter will equip you with the basic knowledge to handle medical emergencies until professional help can be accessed - a crucial skill when you're millions of kilometers from the nearest hospital. Remember, your primary goal is to stabilize the injured party and prevent further harm.

Start with a rapid assessment of the situation. Is the environment safe? Are there continuing dangers like electrical hazards, fire, or depressurization? Your own safety is paramount - a rescuer becoming a casualty helps no one. Once the scene is secure, approach the injured individual. Check for responsiveness - are they conscious? Do they respond to your voice or touch? If unresponsive, check for breathing and pulse. In the low-gravity environment of a spacecraft, checking the carotid artery in the neck is often the most effective method.

If the individual is unresponsive and not breathing, begin CPR immediately. The principles of CPR remain the same, even in microgravity, though the practical application can be challenging. Secure yourself to a stable surface to provide leverage for chest compressions. Use a standard compression-to-breath ratio of 30:2. Remember that exhalation in space requires conscious effort due to the lack of atmospheric pressure, so rescue breaths must be delivered with more force than on Earth.

For a conscious individual, assess for injuries. Bleeding is a serious concern in the pressurized, oxygen-rich environment of a spacecraft. Direct pressure is the first line of defense. Apply firm pressure to the wound using a sterile dressing, if available, or any clean cloth. Elevate the injured limb, if possible, to reduce blood flow. If bleeding continues, consider using a pressure bandage. Tourniquets should be used only as a last resort due to the risk of tissue damage.

Burns can also occur from various sources, including radiation exposure, chemical spills, and electrical malfunctions. For thermal burns, immediately cool the affected area with cool (not ice cold) water for at least 10 minutes. Do not remove any adhered clothing. Cover the burn loosely with a sterile dressing. For chemical burns, flush the area with copious amounts of water for at least 20 minutes, ensuring the runoff doesn't contaminate other areas of the spacecraft or yourself.

Fractures are another potential hazard in the confined spaces of a spacecraft. Immobilize the injured limb using whatever materials are available - splints can be improvised from rolled-up blankets, sturdy plastic, or even spare parts secured with tape or bandages. The goal is to prevent further movement and minimize pain. Do not attempt to realign the bone. Pain management can be achieved with over-the-counter analgesics if available, following the dosage instructions carefully.

In the event of eye injuries, such as foreign objects or chemical splashes, irrigate the eye with sterile saline solution or clean water. If a foreign object is embedded, do not attempt to remove it. Cover the eye with a clean dressing and seek immediate medical assistance.

Hypothermia and hyperthermia can also occur due to malfunctions in the spacecraft's environmental control system. For hypothermia, warm the individual slowly. Remove wet clothing and replace it with dry blankets or clothing. Offer warm liquids if the person is conscious and able to swallow. For hyperthermia, move the individual to a cooler environment, remove excess clothing, and apply cool compresses to the skin. Encourage them to drink cool fluids.

Finally, accurate record-keeping is crucial. Note the time of the incident, the nature of the injury, the treatment provided, and the individual's response to treatment. This information will be invaluable to medical personnel once they can take over care. This log should include vital signs like pulse rate, respiration rate, and any changes in the individual's condition.

Maintaining a well-stocked first aid kit is also essential. It should include a variety of bandages, sterile dressings, antiseptic wipes, pain relievers, antihistamines, anti-nausea medication, scissors, tweezers, and a thermometer. Regularly check the expiration dates of medications and replace them as needed. Familiarize yourself with the contents of the kit and the proper use of each item. A comprehensive first aid manual should also be readily available and consulted regularly. Remember, in the vastness of space, you are the first responder. Your quick thinking and decisive action can be the difference between a minor incident and a major catastrophe.

6.5 Distress Signals

A distress signal is a last resort, a desperate plea for help when all other systems have failed. Understanding how to use them effectively could mean the difference between life and death. This chapter will cover the intricacies of distress beacons, emergency transponders, and the universal distress protocols recognized across the galaxy. We'll delve into the specific signal patterns, the power sources behind these vital tools, and the strategies for maximizing their reach in the cold void of space.

Distress beacons operate on a variety of frequencies, each designed for specific situations and ranges. The standard short-range beacon transmits a pulsed signal on the commonly monitored emergency band. This is your go-to signal for nearby vessels or planetary outposts. Medium-range beacons utilize subspace frequencies, allowing them to punch through interference and reach significantly farther. These are crucial when operating in nebulae or regions with high electromagnetic activity. For extreme long-range communication, the hyperwave distress beacon is employed. This powerful device transmits a highly focused beam capable of traversing vast interstellar distances, but requires a significant power draw. Understanding the strengths and limitations of each beacon type is paramount to efficient distress signaling.

Emergency transponders offer a more sophisticated approach to distress calls. Unlike beacons, which broadcast a generic SOS, transponders can transmit detailed information about your situation. This includes the nature of the emergency, your vessel's identification, and precise location coordinates. Modern transponders even incorporate automated diagnostic systems, capable of transmitting critical ship data, like hull integrity and life support status. This vital information can significantly expedite rescue efforts by providing responders with a clear picture of the situation before they even arrive.

The power source for your distress signaling equipment is just as crucial as the signal itself. Most beacons and transponders utilize a combination of primary and backup power systems. The primary system is usually tied into the ship's main reactor, providing a consistent and reliable energy source. However, in a catastrophic emergency, the main reactor might be compromised. This is where the backup power system becomes essential. Typically consisting of high-capacity energy cells or independent fusion generators, these backup systems are designed to activate automatically upon main power failure, ensuring that your distress signal continues to broadcast.

Maximizing the reach of your distress signal involves understanding the principles of signal propagation and environmental factors. In the vacuum of space, signals travel in a relatively straight line. However, gravitational fields, celestial bodies, and interstellar dust can all affect signal strength and direction. Knowing how to compensate for these factors is crucial. For example, positioning your antenna array to avoid obstruction by the ship's hull can significantly improve transmission range. Similarly, utilizing signal boosters can amplify the signal strength, pushing it further into the depths of space.

Universal distress protocols dictate specific signal patterns and formats to ensure that your message is understood regardless of the recipient's species or technology. The most common protocol involves a repeating sequence of three short pulses, three long pulses, and three short pulses - the familiar SOS signal. This pattern is easily recognizable and transcends language barriers. In addition to the SOS, most distress signals incorporate a unique identification code specific to your vessel. This allows rescuers to pinpoint your ship among potentially hundreds of other vessels in the vicinity.

Beyond the technical aspects, effective distress signaling also involves a degree of psychological preparedness. Panic can cloud judgment and lead to mistakes. Regular drills and simulations can help instill a sense of calm and procedural memory, ensuring that you can react quickly and efficiently under pressure. Knowing the protocols by heart, understanding the functionalities of your equipment, and maintaining a clear head are just as crucial as the technical aspects of distress signaling.

Furthermore, the content of your transmitted message, if possible, should be concise and informative. Include your ship's name and registry, the nature of the emergency, your current location, the number of crew members aboard, and any immediate needs, such as medical supplies or engineering assistance. Brevity is key, as every second counts in an emergency situation. Transmitting unnecessary information can clog the communication channels and delay the arrival of help.

Finally, don't give up hope. Even in the most dire circumstances, the persistent transmission of a distress signal can significantly increase your chances of rescue. The vastness of space may seem isolating, but remember that there are others out there, listening, ready to respond.

A well-prepared engineer understands not just the mechanics of their ship, but also the importance of clear communication in times of crisis. Distress signals are not just a technical procedure; they are a lifeline, a testament to the enduring human spirit in the face of adversity.

7 Chapter 7: Tools of the Trade

Every aspiring rocket mechanic needs a reliable toolkit. Working on interstellar vehicles requires specialized equipment capable of handling the unique challenges of space. This chapter will familiarize you with three essential tools: the laser wrench, the plasma torch, and the anti-gravity clamp. Understanding their operation and maintenance is crucial for any aspiring space engineer.

The laser wrench isn't your average wrench. Forget about applying brute force; this tool utilizes focused laser energy to loosen and tighten bolts with pinpoint accuracy. Different laser settings allow for varying degrees of torque, ensuring that delicate components aren't overstressed while still providing enough power to secure critical joints. Regular calibration is essential to maintain its precision. Look for slight fluctuations in the laser beam's intensity or any unusual humming sounds during operation. These could indicate a power supply issue or a misalignment in the focusing lenses. Cleaning the lens with a specialized microfiber cloth designed for optical equipment is a routine maintenance task. Furthermore, ensure the power pack is fully charged before each use to prevent interruptions during critical repairs.

Next, we have the plasma torch, indispensable for cutting and welding in the vacuum of space. Generating a superheated stream of ionized gas, this tool can melt through even the toughest alloys used in rocket construction. Controlling the gas flow and temperature is key to achieving a clean cut or a strong weld. A wavering or sputtering plasma stream indicates potential issues with the gas supply or the internal electrodes. Regularly check the gas lines for leaks and ensure the connections are secure. The electrodes, responsible for generating the plasma, degrade over time and require periodic replacement. Learning to recognize the signs of wear, like inconsistent arc formation or a change in the plasma's color, will prevent mid-repair failures.

Finally, consider the anti-gravity clamp, an ingenious device that allows you to manipulate heavy components in the weightless environment of space. By generating a localized gravitational field, it can hold and move objects as if they were on Earth. This is crucial when working on large engine parts or sections of the hull. The power source for the anti-gravity clamp is its most critical component. A fluctuating gravitational field indicates a potential power drain or a malfunctioning energy regulator. Regularly inspect the power cables for any signs of fraying or damage. The clamp's internal sensors, responsible for maintaining the gravitational field's stability, also require periodic calibration. A drifting or vibrating object held by the clamp suggests a sensor malfunction. Mastering these tools is a significant step towards becoming a proficient rocket repair technician. Each tool possesses unique capabilities and requires specific maintenance procedures. Regular upkeep and a thorough understanding of their operation will ensure you're ready to tackle any repair challenge in the vastness of space. Remember, precision and care are paramount when working on complex interstellar vehicles. The safety of the crew and the success of the mission rely on the meticulous work of skilled technicians like you. Therefore, dedicate yourself to learning the intricacies of these tools, and you'll be well on your way to a successful career in interstellar vehicle maintenance. Practice makes perfect, and familiarizing yourself with the feel and function of these tools is vital. Simulate repairs in a controlled environment before tackling real-world scenarios. This will build your confidence and prepare you for the demands of working on a rocket in the challenging environment of space. Pay close attention to the nuances of each tool. The subtle hum of a laser wrench, the steady hiss of a plasma torch, and the quiet hum of the anti-gravity clamp's generator all provide valuable feedback about their operational status. Understanding these nuances can help you diagnose problems early on and prevent costly repairs or even catastrophic failures. Furthermore, always consult the manufacturer's manuals for detailed instructions and troubleshooting tips. These manuals are invaluable resources and often contain diagrams, specifications, and step-by-step guides for maintaining and repairing each tool. Keep your toolkit organized and easily accessible. In the confined space of a spaceship, every second counts during an emergency repair. A well-organized toolkit will allow you to quickly locate the necessary tools and complete the repair efficiently. Invest in high-quality tool storage containers designed for zero-gravity environments. These specialized containers will prevent tools from floating away and ensure they are readily available when needed. Finally, never underestimate the importance of safety protocols. Working with high-powered tools in the hazardous environment of space requires strict adherence to safety procedures. Always wear appropriate protective gear, including eye protection, gloves, and a specialized spacesuit designed for repair work. Double-check all connections and power settings before activating any tool. A moment of carelessness can lead to serious injury or damage to the spacecraft. Remember, a well-maintained toolkit and a meticulous approach are the hallmarks of a skilled rocket repair technician. By dedicating yourself to learning the intricacies of these essential tools, you'll be well-equipped to keep interstellar vehicles running smoothly and ensure the success of future space missions.

7.1 Laser Wrenches

Laser wrenches are indispensable tools for any aspiring rocket repair technician. They provide the precision and control necessary for working with the delicate and often complex components found in interstellar vehicles. Unlike traditional mechanical wrenches, laser wrenches utilize focused beams of light to apply torque, offering several advantages in the unique environment of space. This chapter will delve into the intricacies of laser wrench operation, maintenance, and specialized applications, equipping you with the knowledge to wield this essential tool effectively.

One of the primary advantages of laser wrenches is their ability to apply precise amounts of torque. This is crucial when working with sensitive components like sensor arrays or navigation equipment, where over-tightening could lead to irreparable damage. Laser wrenches allow for micro-adjustments, ensuring that each bolt and fastener is secured with the correct amount of force. This level of control is simply not achievable with conventional tools. Furthermore, the non-contact nature of laser wrenches minimizes the risk of scratching or marring delicate surfaces, a significant concern when working with expensive and often irreplaceable spacecraft components. Imagine tightening a bolt on a warp drive nacelle - the precision of a laser wrench is paramount.

Different types of laser wrenches are suited for various tasks. Low-power laser wrenches are ideal for fine adjustments and working with smaller components. These wrenches emit a highly focused beam of low-intensity light, providing the delicate touch needed for intricate work. High-power laser wrenches, on the other hand, are designed for heavier-duty applications, such as securing structural elements or working with larger engine parts. Their more intense beams deliver significantly greater torque, allowing technicians to quickly and efficiently tighten large fasteners. Understanding the power output and application of each type of laser wrench is crucial for selecting the right tool for the job.

Calibration and maintenance are essential for ensuring the continued accuracy and effectiveness of your laser wrench. Regularly calibrating the wrench ensures that the torque applied matches the settings displayed on the tool's interface. This process involves using a calibrated torque sensor to verify the wrench's output at various power levels. Neglecting calibration can lead to over\item or under-tightening, potentially compromising the integrity of the spacecraft. Maintaining the lens of the laser wrench is equally important. A clean lens ensures a focused beam, maximizing the wrench's efficiency and preventing damage to the target component. Dust or debris on the lens can scatter the beam, reducing its effectiveness and potentially causing overheating.

Beyond their primary function of tightening and loosening fasteners, laser wrenches have several specialized applications in rocket repair. They can be used to activate certain types of micro-mechanisms within spacecraft systems, utilizing the precise energy delivery of the laser beam to trigger specific functions. In emergency situations, a laser wrench can even be used to cauterize damaged wires or seal small breaches in the hull, providing a temporary fix until a more permanent repair can be made. This versatility makes the laser wrench an invaluable tool in the unpredictable environment of space travel.

Safety precautions are paramount when operating a laser wrench. The highly focused beam of light can be hazardous to the eyes, so always wear appropriate eye protection. Specialized laser safety goggles designed to filter the specific wavelength of the wrench's beam are essential. Never point the wrench at yourself or others, and always ensure the area is clear before activating the tool. Accidental exposure to the beam can cause serious eye damage or burns. Furthermore, be mindful of the wrench's heat dissipation system. High-power laser wrenches can generate significant heat during operation, and proper ventilation is essential to prevent overheating. Regularly inspect the heat sinks and cooling vents to ensure they are free from obstructions.

Mastering the use of a laser wrench requires practice and a thorough understanding of its capabilities. Start by practicing on simple components before moving on to more complex systems. Familiarize yourself with the different power settings and their corresponding torque outputs. Learn how to adjust the focus of the beam for different applications, and understand the importance of maintaining a steady hand during operation. With experience, you will develop the precision and control necessary to wield this versatile tool with confidence. Remember, a skilled technician knows not just how to use a tool, but also when and why to use it. The laser wrench is a powerful asset in the arsenal of any rocket repair technician, offering the precision and control necessary to keep interstellar vehicles in optimal condition.

7.2 Plasma Torch

The plasma torch isn't your average welding tool. It's a specialized piece of equipment designed to operate in the challenging environment of space, capable of performing both cutting and welding tasks with remarkable precision. Understanding its functionality and safe operation is paramount for any aspiring rocket repair technician. Unlike common earth-bound torches relying on combustion, the plasma torch harnesses the power of superheated, ionized gas - plasma - to achieve incredibly high temperatures. This plasma stream, reaching temperatures exceeding 20,000 degrees Celsius, can melt even the most robust spacecraft alloys with ease.

At the heart of the plasma torch lies the plasma arc. This arc is generated between an electrode within the torch and the workpiece itself. A high-voltage current ionizes a flow of gas, typically argon or a mixture of argon and hydrogen, creating the plasma. This ionized gas, now electrically conductive, completes the circuit, allowing the current to flow and maintain the intensely hot plasma arc. The gas, forced through a constricted nozzle, emerges as a focused, high-velocity jet of plasma, delivering concentrated heat to a precise point on the spacecraft hull. This focused heat allows for exceptionally clean and precise cuts, minimizing the risk of collateral damage to surrounding components. When used for welding, the plasma torch's intense heat quickly melts the base metal, creating a strong, durable weld. The controlled nature of the plasma arc prevents excessive heat spread, minimizing distortion and maintaining the structural integrity of the spacecraft's hull. The torch's versatility is further enhanced by the ability to adjust the gas flow rate and the current intensity, allowing technicians to fine-tune the heat output for different materials and applications. Welding thin sheets of aluminum requires a lower setting than joining thick sections of steel alloy, for example.

Operating a plasma torch in the vacuum of space presents unique challenges. The absence of atmospheric pressure affects the plasma arc's stability and requires precise control of the gas flow. Specialized nozzles designed for vacuum operation are essential to maintain a focused plasma stream. Furthermore, the lack of an atmosphere eliminates convective cooling, meaning the torch itself heats up much faster. Effective cooling systems, typically integrated within the torch's design, are crucial to prevent overheating and ensure the longevity of the tool. These systems often involve circulating coolant through channels within the torch body, dissipating the excess heat generated during operation.

Safety protocols are paramount when using a plasma torch. The intense heat and ultraviolet radiation emitted by the plasma arc pose significant hazards. Technicians must wear specialized protective gear, including welding helmets with darkened visors and heat-resistant gloves, to prevent burns and eye damage. The intense brightness of the arc necessitates eye protection with specific shade ratings, capable of filtering out harmful UV radiation. Adequate ventilation is also crucial when working with a plasma torch, as the process generates potentially harmful fumes and gases, even in the vacuum of space. Specialized vacuum exhaust systems are employed to capture these byproducts and ensure a safe working environment.

Beyond the immediate safety concerns, proper handling and maintenance of the plasma torch are vital for its continued performance. Regular cleaning of the nozzle is essential to prevent clogging and maintain a consistent plasma flow. Inspecting the electrode for wear and tear is equally important, as a worn electrode can compromise the stability of the plasma arc and reduce the torch's effectiveness. Technicians should also be familiar with the torch's power supply and control systems. Understanding how to adjust the current and gas flow is essential for achieving optimal welding and cutting results. Troubleshooting common issues, such as erratic arc behavior or insufficient heat output, requires a deep understanding of the torch's operating principles. A skilled technician should be able to diagnose and rectify these problems quickly and efficiently.

Furthermore, the specific gas used with the plasma torch plays a crucial role in determining the quality of the weld or cut. Argon, with its inert properties, is commonly used for welding aluminum and other reactive metals. Adding hydrogen to the argon mix increases the plasma arc's temperature and improves cutting efficiency, particularly for thicker materials. Selecting the appropriate gas mixture is essential for achieving the desired results and preventing material contamination. Understanding the nuances of these different gas mixtures and their impact on the plasma arc is a key aspect of mastering the plasma torch.

Finally, the plasma torch's versatility extends beyond simple cutting and welding. With specialized attachments and techniques, it can be used for gouging, which involves removing material from a groove or cavity, and for thermal spraying, a process that applies a protective coating to a surface. Mastering these advanced techniques significantly expands the technician's repertoire and enhances their ability to tackle complex repair tasks. These specialized applications often require specific nozzle designs and gas mixtures, further highlighting the importance of a thorough understanding of the plasma torch's capabilities.

7.3 Anti-Gravity Clamp

The anti-gravity clamp isn't some futuristic gadget from a sci-fi movie; it's a fundamental tool for any aspiring rocket technician. Imagine trying to wrestle a multi-ton engine component in the weightlessness of space. Without the proper equipment, you'd be like a leaf in a hurricane, tossed around uncontrollably. The anti-gravity clamp provides the crucial ability to manipulate heavy objects with precision and control, making complex repairs and maintenance possible in zero-gravity environments.

These clamps operate on principles distinct from those of artificial gravity systems found within a spacecraft. Instead of generating a gravitational field, they utilize a combination of powerful magnetic fields and focused energy pulses to counteract the effects of inertia. Think of it like creating a localized "anchor point" in space. This anchor allows technicians to manipulate objects as if they were working under normal gravity, offering stability and precision during complex procedures. Different models use varying combinations of these technologies, each with its own advantages and limitations.

One common type utilizes high-powered electromagnets. These create a strong magnetic field that can interact with the metallic components of a spacecraft or engine. By adjusting the strength and direction of this magnetic field, technicians can lift, move, and secure heavy objects. This type is particularly effective for ferrous metals, but its efficacy diminishes with non-magnetic materials. For such instances, alternative clamp designs are necessary.

Another type of anti-gravity clamp employs what is known as a "gravitic lens." This device focuses a specific type of energy, creating a localized distortion in the fabric of spacetime. This distortion, although minute, creates a small, controlled gravitational pull. While incredibly complex to manufacture and calibrate, gravitic lens clamps offer the advantage of working on any material, regardless of its magnetic properties. They are particularly useful for manipulating delicate components or those made of composite materials. The specific operating principles behind gravitic lens technology are quite advanced, but understanding the basic concept of creating a localized gravitational field is essential for any aspiring rocket technician.

Choosing the right clamp for a particular task is paramount. Factors such as the weight and material of the object, the precision required, and the surrounding environment all play a role. For instance, working near sensitive electronic equipment might necessitate the use of a gravitic lens clamp to avoid magnetic interference. Conversely, manipulating large sections of the hull might require the raw power of a multi-electromagnet system. Understanding these nuances is what separates a competent technician from a true expert.

Maintenance of these clamps is also crucial. Electromagnetic clamps require regular calibration to ensure the magnets maintain their strength and stability. The focusing lenses of gravitic clamps require meticulous cleaning and alignment to maintain their effectiveness. Neglecting these maintenance procedures can lead to catastrophic failure, potentially endangering both the technician and the spacecraft. Imagine attempting a delicate engine repair only to have your clamp malfunction, sending a critical component spiraling into the depths of space. Regular maintenance isn't just a recommendation; it's a necessity.

Proper training is indispensable for operating an anti-gravity clamp effectively and safely. These aren't simple hand tools; they're complex pieces of equipment that require a thorough understanding of their operating principles. Certification programs exist to provide technicians with the necessary knowledge and practical experience. These programs cover everything from basic operation to advanced troubleshooting and emergency procedures. Imagine activating a clamp without properly securing it to the spacecraft. You, along with the object you're attempting to manipulate, would be propelled in the opposite direction, drifting helplessly into the void.

Beyond the practical applications, understanding the underlying physics of these devices is crucial. A deep knowledge of magnetism, gravity, and energy manipulation allows technicians to troubleshoot problems effectively and adapt to unexpected situations. This knowledge is also invaluable for developing new and improved clamp technologies, pushing the boundaries of what's possible in space exploration. The field of rocket repair is constantly evolving, and a strong foundation in the underlying scientific principles is essential for staying at the forefront of this dynamic industry. From basic repairs to complex overhauls, the anti-gravity clamp is an indispensable tool in the arsenal of any aspiring rocket technician. Mastering its use is a key step on the path to becoming a true expert in interstellar vehicle maintenance.

www.ingramcontent.com/pod-product-compliance
Lightning Source LLC
LaVergne TN
LVHW022008200926
848481LV00007B/2283

* 9 7 9 8 3 4 8 3 3 9 8 1 4 *